How the Human Genome Works

Edwin H. McConkey, Ph.D.
Professor Emeritus
Department of Molecular, Cellular, and Developmental Biology
University of Colorado
Boulder, Colorado

JONES AND BARTLETT PUBLISHERS
Sudbury, Massachusetts
BOSTON TORONTO LONDON SINGAPORE

World Headquarters
Jones and Bartlett Publishers
40 Tall Pine Drive
Sudbury, MA 01776
978-443-5000
info@jbpub.com
www.jbpub.com

Jones and Bartlett Publishers Canada
2406 Nikanna Road
Mississauga, ON L5C 2W6
CANADA

Jones and Bartlett Publishers International
Barb House, Barb Mews
London W6 7PA
UK

Executive Publisher: Christopher Davis
Production Manager: Amy Rose
Associate Production Editor: Renée Sekerak
Editoral Assistant: Kathy Richardson
Manufacturing Buyer: Therese Bräuer
Composition and Art Creation: Dartmouth Publishing, Inc.
Printing and Binding: D.B. Hess
Cover Printing: D.B. Hess
Cover Image: © Ryan McVay/Photodisc/Getty Images

Library of Congress Cataloging-in-Publication Data
McConkey, Edwin H.
How the human genome works / Edwin H. McConkey.
p. ; cm.
Includes index.
ISBN 0-7637-2384-3
1. Human genome. 2. Medical genetics. [DNLM: 1. Genome, Human. 2. Genetic Diseases, Inborn. 3. Genetic Predisposition to Disease. 4. Genetics, Medical. 5. Mutation—genetics. QH 447 M478h 2004] I. Title.
QH447.M353 2004
611'.01816—dc22

2003017746

Printed in the United States of America
08 07 06 05 04 10 9 8 7 6 5 4 3 2 1

Contents

Preface

How the Human Genome Works is a brief summary of basic facts about human genes, how they are expressed, how mutations lead to simple and complex disorders, and how the rapid advance in our understanding of the human genome is impacting the practice of medicine. This book will be useful for people in the health sciences at all levels, from students to established professionals, who want to update their knowledge about human genetics without making a major time commitment. *How the Human Genome Works* will also be a good supplementary text for a variety of college courses where the major text does not cover human genetics in sufficient depth to meet the instructors' goals.

This book assumes that you already have had an introductory college biology course or that you at least know a few basic facts about genes, DNA, proteins, etc. If you are not sure whether this book is right for you, take the following True or False quiz.

1. Mendel's Laws were written 2,500 years ago by Middle Eastern mystics living in caves.
2. Watson and Crick were a pair of comedians who appeared in numerous movies during the 1940s and 1950s.
3. ATP is a prominent telecommunications company.
4. Different chromosomes have distinctive colors; X chromosomes are pink and Y chromosomes are blue.
5. Meiosis is a sexually transmitted disease.

If you spent more than a couple of milliseconds considering the possibility that any of the above statements were true, you may not be ready for this book! If you scored 100% on the test, you could then take a look at the Review sections at the end of any of the chapters. If you think you have read all that basic stuff about genes and gene expression in Lecture 1 too many times already, that's no problem—just skip whatever is boringly familiar and move on to the part you need to learn or re-learn. Chances are, you'll find a lot here that will be useful to you.

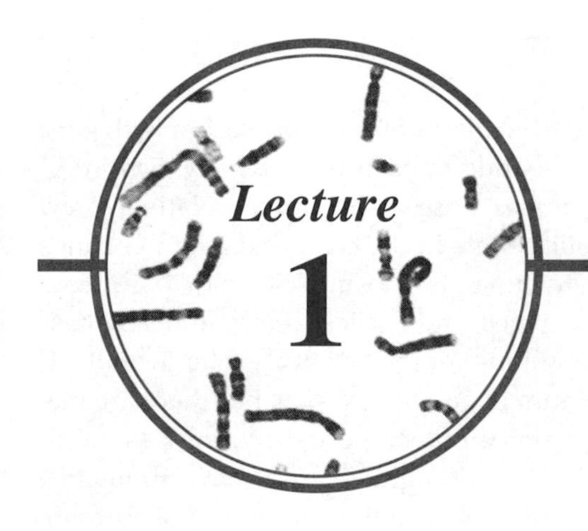

Natural History of the Human Genome

We are living in a time of breathtaking progress in human genetics. This is the genomics era—the time when the DNA sequence of the human genome has been almost completely determined, the time when the functions of thousands of human genes in health and disease are being analyzed. It is the time when the study of small variations in many genes will lead to individualized medicine, the time when the genetic basis of most congenital abnormalities will be revealed, and the time when comparison of human and other primate genomes will reveal the genetic basis of human uniqueness. This is a time of endless excitement for those who participate in the discovery process, either directly in research or indirectly by application of the torrent of new knowledge in the health professions.

GENERAL ASPECTS OF THE GENOME

The word *genome* refers to the entire DNA content of a species, including the genes and all the rest of the DNA. However, *genome* is more than an expression for an amount of DNA; when we speak of a genome, we are really thinking in terms of information content. Humans have the standard mammalian DNA content, which is approximately 3 billion base pairs per germ cell (the *haploid genome*), or 6 billion base pairs in most somatic cells (the *diploid genome*). The latter is equivalent to about six picograms of DNA, which is approximately one-trillionth the weight of a teaspoon of salt. Elephants and mice have roughly the same size genomes as humans, so the fact that we can communicate in hundreds of written and spoken languages, design computers, send rocket ships deep into space, and engage in all sorts of cognitive feats that no other mammal can achieve must be a consequence of a relatively small fraction of our DNA. Identifying the genetic basis for our uniquely human anatomy, physiology, behavior, and cognitive abilities will be one of the greatest scientific adventures of the 21st century. But first, we must complete the catalog of human genes and learn how each of them functions in normal development and metabolism. That information, supplemented by a vast catalog of normal and abnormal genetic variants, will profoundly influence clinical medicine.

DNA was identified as the hereditary material in 1944 when Avery, MacLeod, and McCarty showed that a bacterial phenotype could be changed by treating one strain of cells with DNA, but not protein or RNA, from another strain. The double-helical structure of DNA with the base pairs A-T and G-C between strands was deduced by Watson and Crick in 1953, thus providing a molecular model for information coding and for replication of genetic information.

Long before DNA was recognized as the hereditary material, chromosomes were studied by light microscopy, taking advantage of the fact that at metaphase, the midpoint of mitotic cell division, chromosomes are highly condensed and are physically separate from one another. In the early days of cytogenetics, there was a lot of confusion about the total number of human chromosomes. It wasn't easy to prepare a *metaphase spread* (a microscope slide containing chromosomes spread out so that each chromosome from a single cell is separate from the others), and the chromosomes would appear as dark blobs with no substructure, often overlapping. A well-known cytologist named Painter worked with spermatocytes and concluded in 1923 that humans have 48 chromosomes. No one had a good reason to challenge his data for about 30 years. Eventually, a method for swelling cells in hypotonic (low ionic strength) solutions was devised, and others discovered drugs that could block dividing cells in culture at metaphase, thus making it easier to collect more experimental material. These two technical advances vastly improved chromosome preparations; using them in 1956, scientists obtained convincing evidence that the correct diploid number was 46. Our diploid genome is organized into *23 pairs of chromosomes*, which range in size from 45 million base pairs to approximately 280 million base pairs. There are 22 pairs of *autosomes* and one pair of *sex chromosomes*.

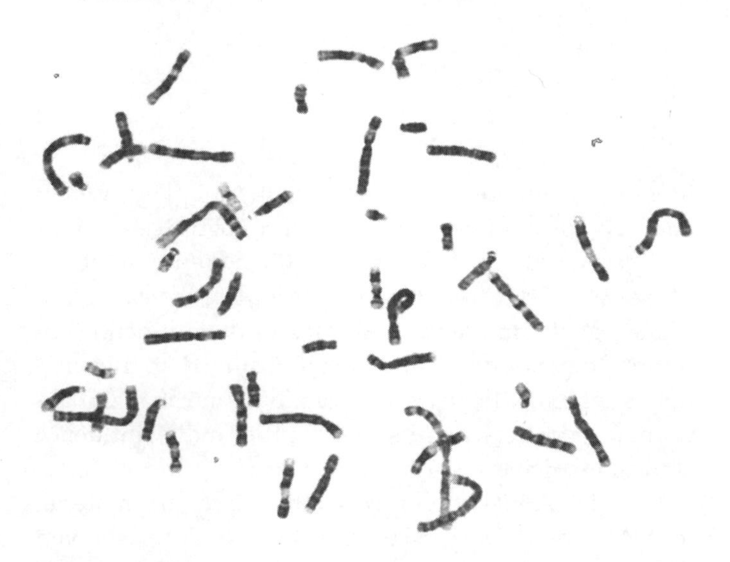

Figure 1-1 A metaphase spread from a normal human male, showing G bands. Courtesy of David Peakman, Reproductive Genetics Inc, Denver.

Another classification scheme used in the 1940s and 1950s was the position of the *centromere*, the chromosome structure to which mitotic spindle fibers attach. Centromeres may be located more-or-less midway between the ends of a chromosome (*metacentric*), noticeably closer to one end than another (*submetacentric*), or quite close to one end (*acrocentric*). Human chromosomes were grouped into seven classes based on size and centromere position.

A major advance in the 1960s was the development of staining methods that revealed alternating light and dark bands within metaphase chromosomes. The most frequently used staining system was devised by a German named Giemsa, and his name is now applied to the bands themselves, which are usually abbreviated *G-bands*. Banding patterns are specific to each chromosome, permitting every chromosome in a metaphase spread to be unequivocally identified. The physical basis for the dark- and light-staining regions is not fully understood, but there is a definite correlation with gene content. Dark G-bands tend to be relatively rich in AT (adenosine-thymine) base pairs, have relatively few genes, and replicate late in the cell cycle. Light G-bands tend to be rich in gunaine-cytosine (GC) base pairs, contain most of the genes, and replicate early in the S phase (DNA replication time) of the cell cycle. The overall GC content of the human genome is 41%, but there are major deviations from the average locally; dark G-bands have about 37% GC, and light G-bands have about 45% GC. We don't yet know whether this aspect of chromosome composition has functional or evolutionary significance.

In 1971, a committee of experts meeting in Paris recommended that the chromosomes be identified by numbers, beginning with #1, the largest, and proceeding down to the smallest. However, the cytological techniques still were a bit uncertain, and they made a little mistake, calling the next-to-smallest #22 and the real smallest chromosome #21. The Paris conference introduced a band numbering system and the letters *p* for the short arm and *q* for the long arm of each chromosome. The sex chromosomes are not numbered; they are called the X and Y chromosomes. Female mammals have two X chromosomes whereas males have one X and one Y chromosome. The X chromosome is medium-sized with an average number of genes. The Y chromosome is quite small; it contains few genes, some of which are necessary for determining maleness, but it is mostly heterochromatin (long arrays of repetitive and highly condensed sequences that don't encode proteins).

The full set of human chromosomes is called the karyotype. One way to describe the normal human karyotype is 46,XY for males or 46,XX for females; you will learn about some abnormal karyotypes in Lecture 2. Analyses of metaphase spreads after G-banding can be converted to neatly arranged diagrams that are called *ideograms*. An ideogram of the human karyotype is presented here.

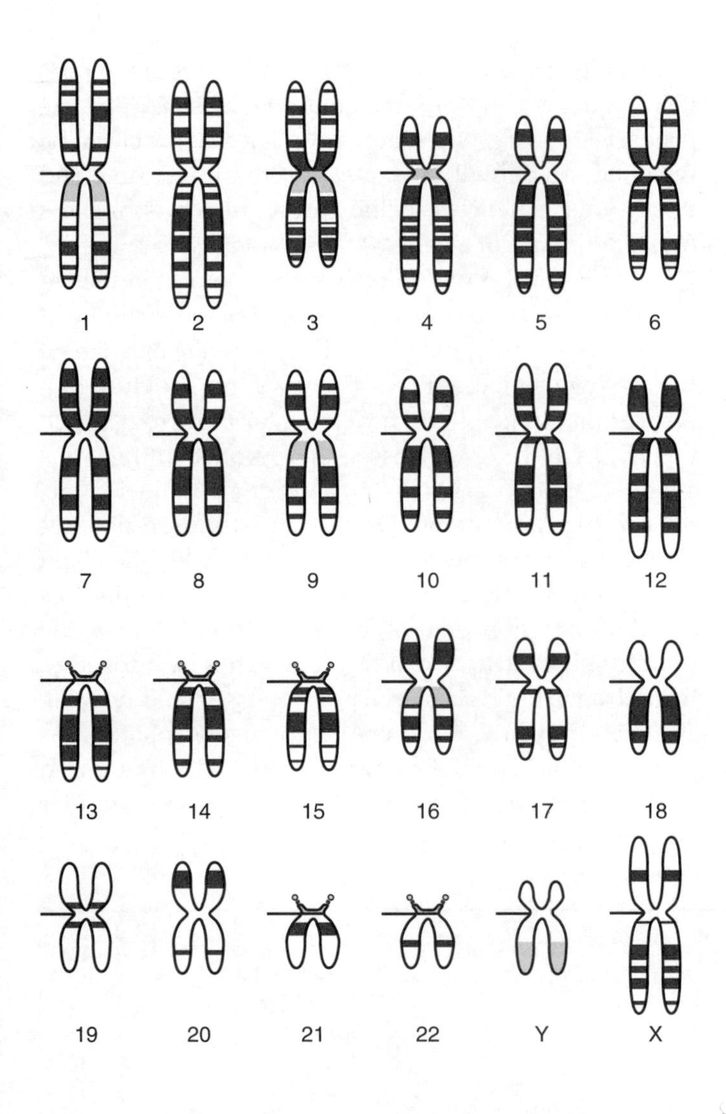

Figure 1-2 Ideogram of human chromosomes, showing bands obtained with various dyes. From Harnden and Klinger, ISCN 1985, p112–113, by permission of S.Karger SG.

During the 1970s and 1980s human genetics was dominated by *gene mapping* studies. One fruitful approach was the creation of *hybrid cells,* containing a complete rodent genome and one or a few human chromosomes. When a specific human gene product or enzyme activity could be identified in such hybrids, it was possible to deduce which chromosome contained the corresponding gene. Then the development of recombinant DNA technology gave us the power to clone human genes, thereby producing unlimited quantities of small genomic pieces in microorganisms. Cloning, when combined with a variety of new physical techniques and traditional genetic methods, tremendously increased the power of gene mapping and analysis.

A crucial advance for genetics in general was the development of DNA sequencing methods in 1977. Until that time, the possibility of extensively investigating the human genome and the molecular basis of human genetic disease had seemed hopeless. By the mid-1980s DNA sequencing had been improved sufficiently that large-scale sequencing projects could be reasonably envisioned. A group of scientists associated with the Department of Energy's biology program, which was aimed at studying the human mutation rate (see Lecture 2), realized that it was now possible to search for mutations at the DNA level. But searching for DNA mutations also requires analysis of the normal DNA sequence. Therein lay the seeds of the *Human Genome Project (HGP),* which was officially begun in the fall of 1990, as a joint effort by the National Institutes of Health, which established the *National Human Genome Research Institute (NHGRI)* and the Department of Energy. The project eventually expanded internationally, with significant contributions coming from the United Kingdom, France, Germany, Japan, and China. Those countries formed the *International Human Genome Sequencing Consortium.*

The central goal of the Human Genome Project was to sequence the entire three billion base pairs in the human haploid genome and identify all the genes. A major progress report was published in February 2001. Approximately 90% of the genome had been sequenced at that time, but a significant fraction of the total sequence was still in the form of pieces only a few thousand base pairs long, which greatly limited its usefulness. For an analogy, imagine that every sentence in this book was written on a separate piece of paper and when you paid the clerk at the bookstore, you were handed a bag with all those pieces in random order. You wouldn't learn much about human genetics that way!

Nearly all of the gaps were filled in the next two years, and the essentially complete sequence of the human genome was announced in April 2003—an appropriate event to celebrate the fiftieth anniversary of the Watson-Crick model of DNA structure! About 99% of the gene-containing regions have been sequenced to an accuracy of 99.99%. There are fewer than 400 gaps remaining, and the average nucleotide is part of an uninterrupted stretch of more than 27 million base pairs. The overall size of the genome is about 3.2 billion base pairs. A surprising result was that the total number of human genes is in the range of 30,000–35,000, much lower than some earlier estimates, and only about twice the number of genes in the fruitfly or the worm.

Completion of the human genome sequence is a major milestone in human genetics, but it certainly does not mean that we know how all the genes function. Analyzing the functions of thousands of genes in health and disease will be a much larger project than sequencing the genome. Plans are already underway to exploit the fundamental sequence data in many ways, which will be described in several contexts in later lectures in this book. For now, it suffices to note that having the complete human genome sequence opens possibilities for understanding human biology and human illness at a level of detail that was unimaginable a few years ago.

WHAT'S IN A GENE?

A gene is defined as a segment of DNA that is transcribed into an RNA copy of one strand of the DNA (a process that will be described later in this lecture). Most genes are regions of DNA that contain the information for the amino acid sequence of a protein, but some genes only encode RNA. Essentially all of the metabolic functions of living cells are mediated by proteins, while other proteins produce the majority of intracellular and extracellular structures. Associated with all genes are regulatory sequences, which are sequences of DNA that are primary binding sites for proteins that determine whether a gene will be expressed at a given time and place. Some geneticists include the regulatory sequences in the definition of a gene.

Figure 1-3 will refresh your memory on the basic aspects of DNA: panels A and B show the structure of the four bases: adenine (A) and thymine (T) or guanine (G) and cytosine (C). Panel C shows the structure of a nucleotide, which is the fundamental unit from which both DNA and RNA are assembled. Note that the terms 5′ and 3′, which designate the ends of a polynucleotide, refer to the carbon atoms in the sugar portion of a nucleotide: The first nucleotide in a polynucleotide chain has a 5′ phosphate at its end, and the last nucleotide has a 3′ hydroxyl group at its free end. Panel D shows the double helical structure of DNA, which is held together by hydrogen bonds between pairs of bases (A-T or G-C). The thick gray lines represent the sugar-phosphate backbone, to which the bases are attached. The figure also shows a nascent RNA strand, which we will consider later in this lecture.

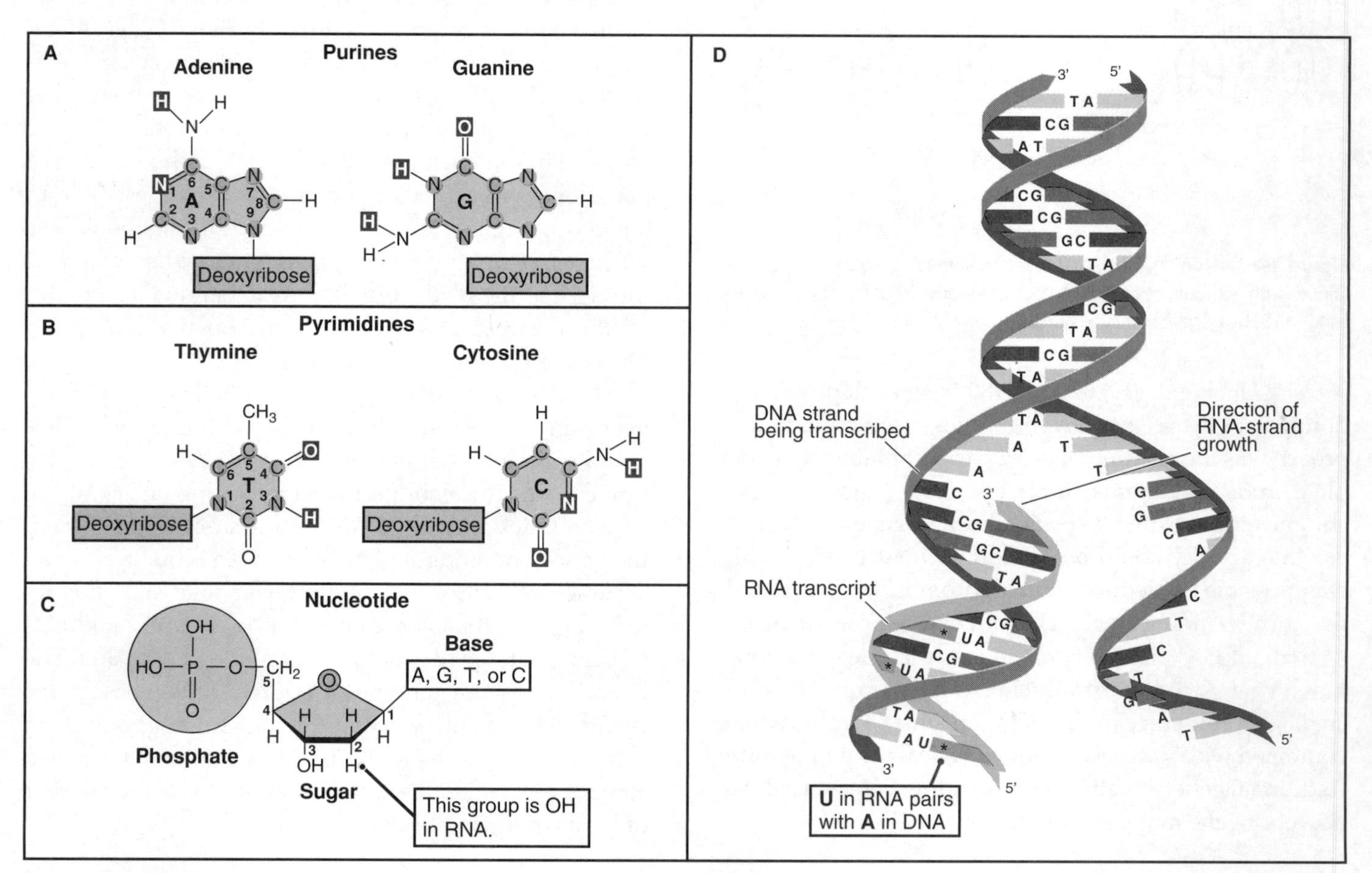

Figure 1-3 The structure of DNA.

Before we delve into gene structure, we should clarify some terminology. The length of a DNA sequence should logically be expressed in nucleotide pairs, because the subunits of DNA (and RNA) are nucleotides, but we usually don't do it that way. For double-stranded DNA, we refer to *base pairs (bp)*, because it is the pairing of the bases between nucleotides that holds the two strands of native DNA together. A thousand bp is a *kilobase* pair, but instead of kbp, we usually drop the p and use the abbreviation *kb*. Similarly, a million base pairs is abbreviated *Mb (megabase)* and a billion base pairs is a *Gb (gigabase)*. Single-stranded DNAs and RNAs have to be described in terms of *bases* (not pairs) or nucleotides (abbreviated *nt*). The "front end" of every DNA and RNA is called 5′ (five prime) and the "tail end" is called 3′ (three prime); those numbers refer to the structure of nucleotides and the way in which they are assembled into DNA or RNA (see Lecture 2).

A typical human gene consists of alternating exons and introns. Exons are DNA segments that will be represented in the mature *messenger RNA (mRNA)* that will be produced when that gene is expressed. Most exons contain the information for a series of *amino acids*, the subunits of proteins. There are also exons at the beginning and end of mRNAs that do not code for amino acids, which may contain various types of regulatory information. Introns are portions of genes that lie between the exons; they are not represented in mRNA. The exon/intron ratio of genes varies widely; a few genes are intronless, whereas in others the introns make up considerably more than 95% of the gene. The function of introns and their evolutionary origins are still not fully understood, but it is generally agreed that having genes constructed of a series of short coding pieces (the exons) allows evolutionary flexibility. The following diagram shows the exon/intron structure of two well-studied genes, which code for the alpha-globin and beta-globin polypeptides of hemoglobin.

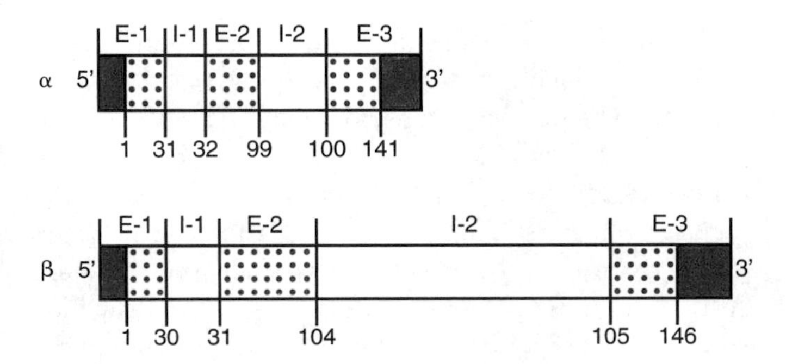

Figure 1-4 The structure of human globin genes. E-1, E-2, and E-3 are exons; I-1 and I-2 are introns. The filled spaces at each end will be included in messenger RNA, but they are not translated. The numbers below the bars indicate the encoded amino acids.

According to the report of the Sequencing Consortium, an average human gene contains 27 kb of DNA. If we multiply 27 kb by 30,000 genes, we learn that genic DNA accounts for 0.8 Gb or roughly one-fourth of the total genome. The Sequencing Consortium has given us some other important numbers about human genes. The average number of exons per gene is approximately 8 (so the average number of introns must be 7). The mean size for exons is 145 bp, and the mean size for introns is 3,365 bp. You can easily calculate that exons account for less than 5% of an average gene. The total amount of DNA in coding exons is an average of 1,340 bp per gene, which is enough to produce a protein 447 amino acids long. However, there is great variability in gene size, number of introns, size of encoded protein, and so on. The largest known gene extends over 2.4 Mb, introns at least 30 kb long are known, and some proteins contain more than 3,000 amino acids.

The estimate of 30,000-35,000 genes referred to earlier was made by a computerized investigation of the genomic sequence, looking first for *known genes* (the easy part), then for genes whose presence could be deduced from the existence of probable exons, exon/intron boundary sequences, and other features. There is uncertainty in the number of *predicted genes*, because the computer may miss some genes and it may predict other genes that do not really exist.

One way to detect an unknown gene is by sequence similarity to a known gene. Many protein-coding genes occur as gene families, which are groups of genes having significant sequence similarity. The basic event in the origin of gene families is gene duplication, which occurs occasionally as an error in DNA replication (Lecture 2). When two copies of a gene exist, one of them may mutate in such a way that it encodes a protein with slightly different properties from the original protein. If that difference is advantageous, selection may perpetuate it. Subsequently, various other mistakes during DNA replication and/or recombination may increase the members of the gene family and disperse some of them to distant locations.

The classic examples of gene families in humans are the beta-globin gene cluster on chromosome 11 and the alpha-globin gene cluster on chromosome 16. Notice the pseudogenes in each cluster (indicated by the Greek letter psi). These arise from duplicated genes where one copy has acquired mutations that make it incapable of being expressed. Many other gene families are known, some of which have dozens of members; examples are the actins, myosins, apolipoproteins, histones, and immunoglobulins. When more distant relationships are analyzed, many genes can be classified into superfamilies, which can have hundreds of members.

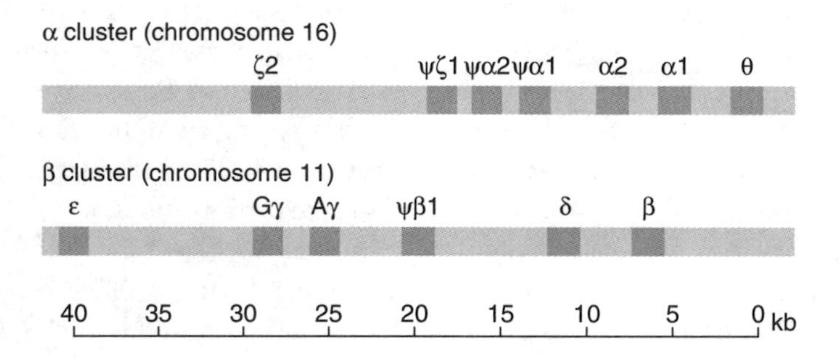

Figure 1-5 The alpha-globin and beta-globin gene clusters.

Let's calculate one other interesting number. What percent of the human genome contains protein-coding information? Multiplying 1,340 bp by 30,000 genes gives us 40,200,000 bp, which is the total coding content. Divide by the overall haploid genome size (3.2 Gb) and the result is that only 1.25% of our genome is protein-coding information. These numbers are approximations, so don't be concerned if you read slightly different ones elsewhere; the important point is that only a small fraction of human DNA encodes proteins.

What is the rest of the genome? We know that 20–25% is occupied by introns, but a large part of the remainder is *intergenic DNA,* and certain crucial portions of that intergenic DNA are *regulatory sequences,* which we will discuss presently. There are also several groups of *genes that do not code for proteins;* the RNA products of those genes take part in many essential steps of cell function, sometimes forming nucleoprotein structures, sometimes directing enzymes to their targets on other RNAs. I will mention specific classes of functional *noncoding RNAs* in various places throughout this book. There are also some portions of the genome that have a structural role. Nevertheless, we are left with no obvious way of explaining why we have so much DNA that is not directly involved with gene structure or function. Some insight into the "excess DNA" problem can be gained by looking at the genome from a different perspective, which we will now do.

REPETITIVE DNA

The DNA in any complex genome can be classified as *single-copy DNA* (sequences present once per genome) and *repetitive or reiterated DNA* (sequences present more than once). Nearly 50% of the human genome is repetitive DNA. A popular term for most repetitive DNA and some of the single-copy DNA that is not part of genes is *junk DNA.* Well, yes, our genomes probably do contain a certain amount of DNA that doesn't have any function at present and can legitimately be considered as junk; but the more we learn about genomes and the control of gene expression, the more we discover functions for DNA that we didn't understand previously. In addition, much DNA that doesn't have any apparent use at present certainly provides materials for evolution of the genome; so if we think of our species as a dynamic entity changing with time, there probably really isn't a lot of junk in the genome.

There are two classes of repetitive DNA: (1) tandemly repetitive sequences, where the repeated sequences are linked together in long series, and (2) dispersed repetitive sequences, where identical or very similar sequences are scattered around the genome, usually with only one copy present at a given site.

Tandemly Repetitive Sequences

A major class of tandemly repeated sequences is centromeric DNA. The most abundant type of centromeric DNA is called *alphoid DNA,* in which the repeat unit is approximately 170 bp long. These units are arrayed in long series that vary from 250 kb to nearly 5 Mb in length, accounting for at least 3% of the genome. Within the series there is some variation in sequence and there is even more variation from chromosome to chromosome. Centromeric DNA organizes the *centromere,* a complex structure containing DNA and several types of proteins, to which the spindle fibers attach during cell division.

A tandemly repeated sequence is found in the telomeres, which are located at the ends of every chromosome. The human telomeric sequence is GGGTTA; it is reiterated from 250 to 1,500 times at different chromosome ends. In recent years, *telomeric DNA* has been the focus of much excitement, as the relationship between telomere shortening and cell senescence was elucidated. Although I don't discuss the details here, the basic idea is that one consequence of the process of DNA replication, which occurs before each cell division, is a shortening of telomeric DNA. If this continues long enough, the telomeric sequences will be eliminated and genes near the telomeres may be damaged. After that, the cell may die or become unable to divide. This is what happens in most normal somatic cells; they have a limited ability to multiply, either in the organism or in cell culture. However, germ cells, stem cells, and various cancer cells contain an enzyme (*telomerase*) that restores the telomeric sequences that would otherwise be lost at each round of DNA replication. Telomerase is an unusual enzyme with both RNA and protein components; the RNA serves as a template for the restoration of missing telomeric repeats. This figure outlines how this process works.

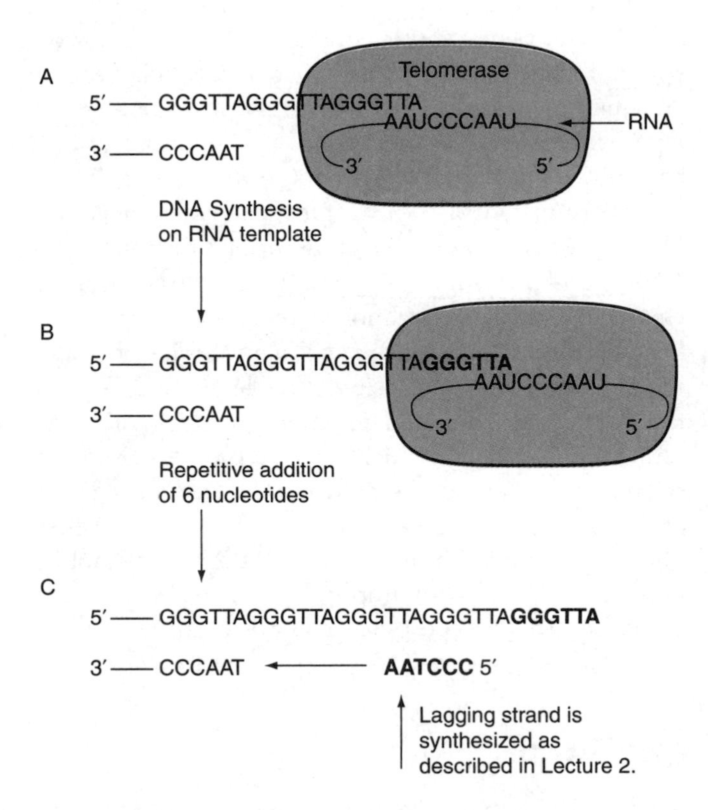

Figure 1-6 Telomere maintenance.

Several other classes of tandemly repeated sequences are genes for noncoding RNAs. One of the best-known contains the genes for *ribosomal RNA*. In humans there are five groups of these genes in the haploid genome, each containing about 60 copies. They are located on chromosomes 13, 14, 15, 21, and 22 in the short arms of those acrocentric chromosomes. These clusters of *rRNA genes*, together with some additional DNA, are called *nucleolus organizers*, because a nucleolus can form at each of them. Nucleoli are ribosome assembly factories. There are more than 80 types of ribosomal proteins. They are synthesized in the cytoplasm and migrate to the nucleus, where they bind to rRNA as it is being synthesized in the nucleolus. A variety of other proteins also accumulate in each nucleolus, where they have roles in the ribosome assembly process. In addition, several types of small, noncoding RNAs participate in ribosome assembly. There is also a cluster of tandemly reiterated genes for *5S rRNA*, another essential component of ribosomes on chromosome 1.

Dispersed Repetitive Sequences

Dispersed repetitive sequences occur singly, rather than in clusters. They are classified into two groups based on size: **LINE** is a acronym for Long **IN**terspersed **E**lements and **SINE** is an acronym for **S**hort **IN**terspersed **E**lements. Both classes of dispersed repetitive sequences are mobile genetic elements called retrotransposons. A fully functional retrotransposon can propagate itself or related sequences as described in the next paragraph. Retrotransposons may have originated as retroviruses, which we will consider in Lectures 4 and 6.

One major class of dispersed repetitive sequences is the LINE-1 or L1 group, of which up to 500,000 copies are present in the human genome, amounting to about 15% of the genome. Most L1 elements are truncated copies of the full-length unit, which is about 5,000 bp long, but there are several thousand full-length L1s. Only 40–50 of them are functionally active; that is, they encode several proteins that are able to cause transposition, either of the L1 unit or some other transposable element. L1 elements have two genes (called ORFs, open reading frames, in Figure 1-7). ORF1 encodes a nucleic acid-binding protein (p40); ORF2 encodes both *reverse transcriptase* (an enzyme that makes a DNA copy, *cDNA*, from mRNA) and an *endonuclease* that makes cuts in the genome where the new cDNA will be inserted.

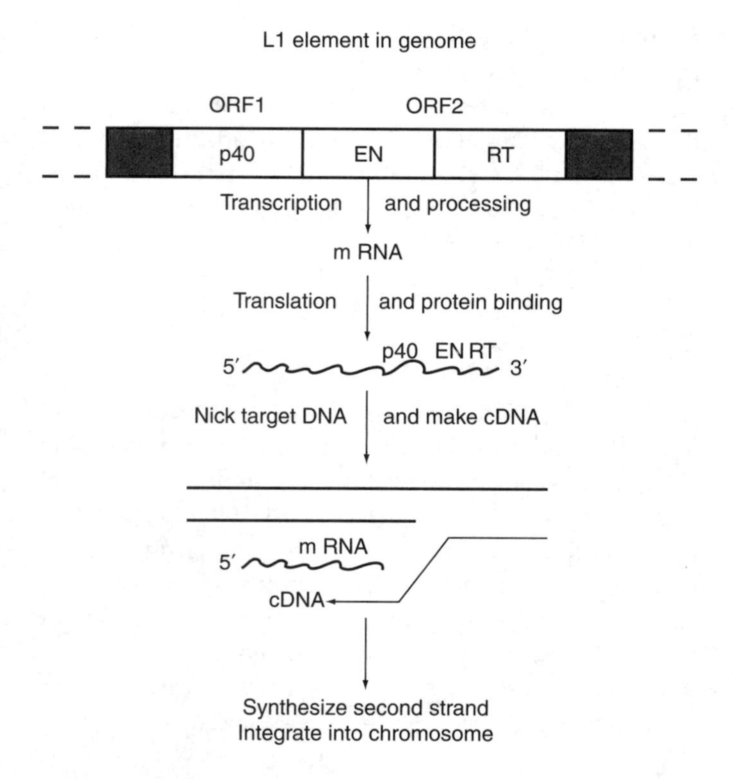

Figure 1-7 Retrotransposition of an L1 element.

When the mRNA from an L1 element has been translated, the proteins it produced usually bind directly to their own mRNA; then that RNA-protein complex moves into the nucleus, where the endonuclease nicks DNA, creating a free end. Reverse transcriptase uses that free end as a primer and makes a DNA copy of the L1 mRNA. Finally, the second strand of cDNA is made and the double-stranded molecule is integrated into the chromosome at the nicked site. We don't yet know why only a portion of an L1 element is the most frequent product of retrotransposition.

It is believed that L1 reverse transcriptase is also responsible for *processed pseudogenes*, which are DNA copies of mRNAs that have been inserted into locations unrelated to the original gene from which the mRNA arose. Processed pseudogenes do not contain introns and are usually not capable of being expressed as polypeptides (but there are a few exceptions), either because they have no regulatory sequences and/or because they contain mutations. Pseudogenes (both processed and ordinary types) are abundant, accounting for 0.5 to 1.0% of the human genome. In chromosome 22, for example, the sequencing team found 134 pseudogenes.

The largest SINEs class is composed of Alu sequences (the name refers to the restriction enzyme Alu I, an endonuclease that cleaves DNA at a specific short sequence, which can be used to cut Alu sequences out of genomic DNA). There are approximately 1 million Alu sequences in the human genome, accounting for 10 to 12% of total DNA. The basic unit is about 300 bp long, but there are many different sequences in the Alu class. They are mostly found between genes and within introns, but rarely are included within messenger RNA. Alu sequences do not encode proteins and thus are not able to move from one location to another by themselves. However, many Alu sequences are transcribed, and there are some short sequences at each end that are similar to those on L1 RNAs, so it is widely believed that enzymes produced by L1 elements participate in *Alu retrotransposition*, although definitive proof has not yet been obtained.

Insertion of a new transposable element into DNA is potentially capable of disrupting the function of a gene, and indeed, more than 30 examples of disease-producing retrotranspositions are known in humans. Furthermore, the presence of so many copies of related sequences creates the possibility for losses and duplications of genomic material during meiosis (as will be described in Lecture 2), which can also produce abnormal phenotypes. Therefore, why don't our genomes get rid of these dangerous pieces of seemingly useless DNA? The answer probably involves evolutionary flexibility. Repetitive DNAs are an important source of genome remodeling. This will become clearer as we discuss mechanisms of genomic change in the next lecture.

CHROMATIN

A linear DNA molecule as long as an average human chromosome (roughly 140 Mb) has a length-to-width ratio so extreme it is difficult to visualize. Analogies with spaghetti and hair long enough to reach a woman's waist are totally inadequate. Imagine a garden hose with an outside diameter of one inch; if it were proportionally as long as an average human chromosome, it would extend for more than 350 miles! In real terms, that 140 Mb chromosome will be more than 47,000 micrometers long, and 46 chromosomes have to be packaged into a nucleus that is no larger than 10 micrometers in diameter in most human cells. The possibilities for entanglement and for accidental breakage are obvious. Compaction of DNA and protection against breakage are achieved via DNA-binding proteins.

DNA in all eukaryotes is associated with a large variety of proteins; the complex is called chromatin. Some of those proteins have a structural role. Many others come and go frequently in connection with gene expression. The basic structural unit of chromatin is the nucleosome, a DNA-protein assemblage consisting of 8 histones and 200 bp of DNA. Histones are basic (positively charged) proteins that come in five main varieties. A *nucleosome core* contains two molecules each of histone 2A, 2B, 3, and 4. They form a disk-shaped particle around which DNA coils approximately two times in such a way that about 150 bp are associated with the nucleosome and another 50 bp lie between one nucleosome and the next. This condenses the DNA filament between six- and seven-fold. Histone 1 attaches to the outside of the DNA-histone octamer complex, probably serving both to stabilize it and to help direct the binding of DNA to the octamer.

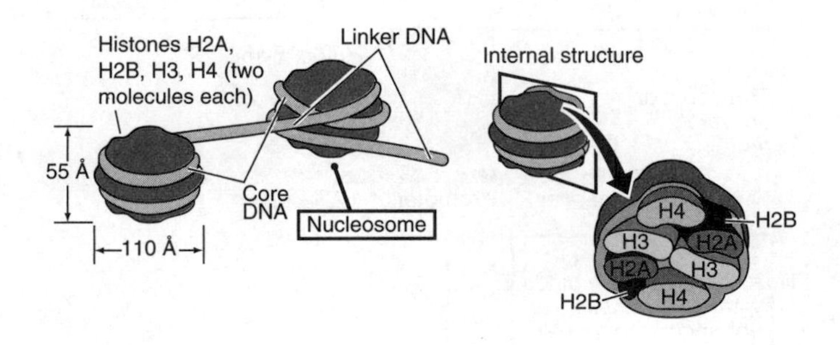

Figure 1-8 Nucleosome structure.

Electron microscope studies on chromatin reveal the existence of the next higher order of folding—*a fiber 30 nanometers (nm) in diameter,* which represents another six- to seven-fold compaction. The exact structure of the 30 nm fibers *in vivo* is still uncertain, because preparation of samples for electron microscopy may distort relationships that exist inside cells. Various structures, such as solenoids, ribbons, and superbeads, have been proposed. An even higher order of chromatin structure is represented by *loops,* extending from a central matrix in the chromosome. The loops apparently consist of 30 nm fibers, or a somewhat more coiled form of them; they range from 30 to 90 kb in length. Microscopy on cells from various organisms indicates that the loops are where active gene expression occurs.

GENE EXPRESSION

Gene expression is the process of transforming a sequence of nucleotides in DNA into a sequence of amino acids in a polypeptide. It requires three major steps: transcription of the gene into RNA, processing of the initial transcript into messenger RNA (mRNA), and translation of the mRNA into a polypeptide. You should already be familiar with the basic concepts and the major classes of macromolecules that mediate those processes; the following brief summary is intended to refresh your memory. As you read these sections, it may help to maintain a clinical perspective: Any reaction in which proteins or RNAs participate is capable of being abolished or modified by mutations in DNA. A given genetic disease may result from aberrations at any step of gene expression.

Transcription

Transcription is the process of copying a segment of DNA in the form of a complementary RNA. RNAs are linear polymers assembled from four nucleotides (nucleoside triphosphates): ATP, UTP, GTP, and CTP. For both DNA and RNA, each new nucleotide in a growing chain is added to the 3′ hydroxyl of the sugar portion of the previous nucleotide with the simultaneous release of the two terminal phosphates from the new nucleotide. RNAs contain the sugar, *ribose,* rather than the *deoxyribose* found in DNA. RNAs also differ from DNAs by using UTP instead of TTP, and in being single-stranded molecules (although they may have small double-stranded regions that arise from internal base pairing). A nascent RNA molecule being synthesized on a DNA template is shown in Figure 1-3D.

Transcription is carried out by RNA polymerases. Bacteria have only one type of RNA polymerase, but in humans, as in all eukaryotes, there are three classes of RNA polymerase: *Pol I* transcribes the genes for the large ribosomal RNAs; *Pol II* transcribes protein-coding genes and some small noncoding RNAs (snRNAs) that have a variety of roles; and *Pol III* transcribes genes for transfer RNA (tRNA), the 5S rRNA, and a variety of other small noncoding RNAs. Each of these enzymes is a complex protein containing several to many polypeptides.

RNA polymerases bind to DNA at promoters, which may be as short as 40 bp, but often are several times that size. Promoters are usually positioned quite close to the beginning (the 5′ end) of the gene. The promoters to which Pol II binds are usually centered about

25 bp upstream of the transcription start site in mammals. They may contain several regions, the most common of which is the *TATA box* (typically the sequence TATAAAA). The affinity of promoters for RNA polymerase varies over several orders of magnitude and control of transcription in human cells requires the presence of other proteins—*transcription factors*—that increase or decrease the probability that RNA polymerase can bind to DNA and initiate an RNA chain. There are a variety of *general transcription factors*, which are proteins distinct from RNA polymerase, but necessary for transcription at most genes. For example, a group of general transcription factors called TFII are required for binding of Pol II to the promoter and establishment of an active *preinitiation complex* that is ready to begin transcription. Binding is usually in the sequence DABpolFEH, where the uppercase letters refer to subunits of TFII. This transcription complex involves several dozen polypeptides.

DNA sequences to which RNA polymerase and its associated regulatory proteins bind are termed cis-regulatory regions, which means that they are located near a particular gene on the same chromosome. Genes that code for regulatory proteins usually have no topological proximity to the genes that they regulate; the proteins are *trans-acting factors*. Cis-regulatory regions also contain binding sites for several different proteins that control expression of the gene in different cell types or at different stages of development. Hundreds of *specific transcription factors* have already been identified, and it is no surprise that mutations in the genes that encode those proteins can be responsible for genetic diseases.

Although the basic preinitiation complex described earlier is competent for transcription, specific transcription factors are usually necessary for the full activity of a gene. The cis-acting sequences to which specific transcription factors bind are known as enhancers. Another term for specific transcription factors is *activators*. Binding of activators to enhancers may increase the frequency of transcription many-fold. Enhancers vary in size from 50 to 1,500 bp and may be located upstream (5′), downstream (3′) or within a gene. Sometimes an enhancer element may be several to many thousand bp away from the gene. Laboratory work has shown that most enhancers can still function if they are moved to a new location (but still nearby on the same chromosome), and their 5′–3′ orientation doesn't matter. This suggests that as long as an activator can bind to the enhancer, it can do whatever it does, within very broad topological limits. The mechanism of action of enhancers is still under investigation and is surely complex, but one basic concept is that when the appropriate proteins have bound to an enhancer sequence, the *configuration of chromatin* in the vicinity of the associated gene is changed. If the enhancer is hundreds of bp (or more) away from the promoter, this configurational change may involve the formation of *chromatin loops*, so that enhancer-bound proteins can interact directly with promoter-bound proteins. Several other possibilities are being studied however. In addition, there is evidence that some enhancer-bound proteins do not directly contact the basal transcription complex, but first interact with another protein known as a *mediator*, which then helps recruit the components of the basal complex to the promoter.

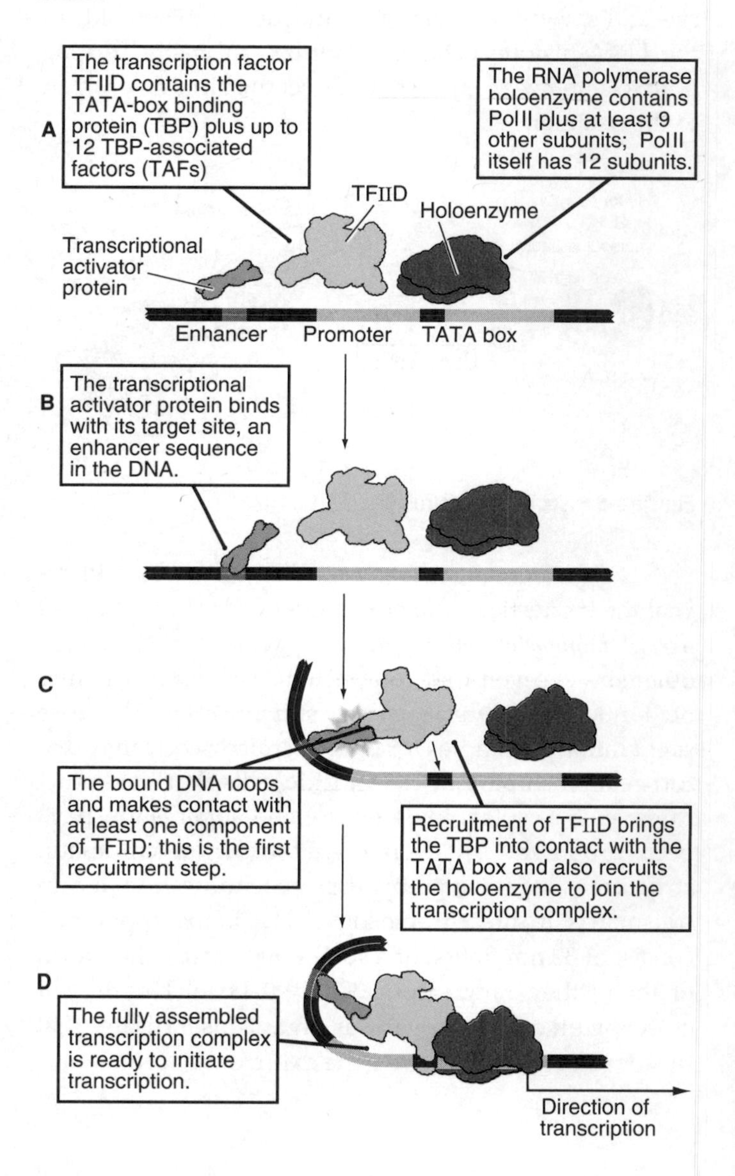

Figure 1-9 Activation of transcription.

For many metabolic processes, there is an equivalent opposite process, and so it is with regulation of transcription: There are cis-acting silencer sequences in DNA that prevent transcription of a specific gene. One hypothesis about silencer function is that when a repressor protein is bound, it causes chromatin to coil up tightly locally and make the promoter of a nearby gene inaccessible to RNA polymerase. However, some DNA elements can be either enhancers or silencers, depending on what molecules are bound to them. A well-known example is the *thyroid hormone response element*, which functions as a silencer when thyroid hormone receptor binds to it without the hormone, but acts as an enhancer when the receptor plus hormone binds to it.

Insulators are DNA sequences that mark boundaries in chromatin beyond which the influence of enhancers and silencers cannot spread. Their mechanism of action is not yet well-defined, but one possibility is that specific proteins bind to insulator DNA and also to some sort of structural components within the nucleus, creating loops of chromatin. The theory also postulates that an enhancer-protein complex in one loop cannot interact with a promoter-protein complex in another loop; however, a convincing molecular model has not yet been developed.

A second function of insulator elements is to prevent the spread of *heterochromatin*—a highly condensed chromatin in which overall transcription is suppressed—into areas where there are active genes. The transcription-suppressing ability of heterochromatin has been known in principle for nearly a century, when the pioneers of genetics of the fruitfly, *Drosophila*, discovered *position effects*. The term refers to the fact that the ability of a gene to be expressed depends on where it is located in the genome. Moreover, there can be *position effect variegation*, which means that the timing and location of gene expression may vary from one tissue or organ to another, depending on the gene's topological relationship to heterochromatin. Position effects are one possible consequence of chromosomal rearrangements (Lecture 2).

Some genes have multiple cis-regulatory elements (promoters, enhancers, etc.) These permit the expression of a gene to vary from one tissue to another, or one developmental stage to another (see Lecture 7), depending on which transcription factors are available to interact with which regulatory element. Another important source of variability in gene expression is in RNA processing, which will be explained in the next section.

RNA Processing

In prokaryotes, the primary transcript of a gene is messenger RNA, but in eukaryotes, that is not the case. As noted earlier in this lecture, nearly all human genes and their primary transcripts contain *introns* (which do not appear in mRNA) between each two *exons* (which do become part of mature mRNA). Introns are also called *intervening sequences*, and exons are called *expressed sequences*. In lower eukaryotes, introns tend to be small and few in number; in mammals (and therefore in humans) however, most genes contain several to many introns, varying in size from a few dozen to thousands of nucleotides. How does the cell process the primary transcript so that the introns are deleted?

Removal of introns and joining of adjacent exons is called RNA splicing. Three types of short nucleotide sequences in the newly synthesized RNA are crucial for splicing: *a consensus sequence at the 5′ end of the intron, a consensus sequence at the 3′ end of the intron, and a branch site about 30 nucleotides from the 3′ end of the intron*, just upstream of a sequence of 8 to 10 pyrimidines (cytosine or uracil). The 5′ consensus sequence almost always includes the *dinucleotide GU* in the intron at the exon/intron boundary, whereas the 3′ consensus sequence almost always includes the *dinucleotide AG* at the intron/exon boundary. The composition of the other parts of the consensus sequences varies considerably. For mammals, one way to express them is:

5′exonAG/GUAAGU-most of intron-YNCURAC-YnNAG/G-3′exon

where Y stands for C or U (pyrimidines), R stands for A or G (purines), N stands for any base, and the slashes indicate the exon/intron boundaries.

As shown in Figure 1-10, the first step in splicing brings an A nucleotide at the branch site into contact with the G nucleotide at the 5′ boundary, where cleavage takes place, freeing the 3′ end of the 5′ exon (called the *donor exon*). The G at the 5′ end of the intron forms a covalent bond to the A at the branch site, forming a *lariat* (a loop with a stem). In the second splicing step, the 3′ OH group of the donor exon attacks the phosphodiester bond at the 3′ end of the intron (the *acceptor splice site*), cleaving it and joining the exons together. These steps are catalyzed by the RNAs themselves, which are said to be functioning as RNA enzymes or *ribozymes*, but they do not act alone.

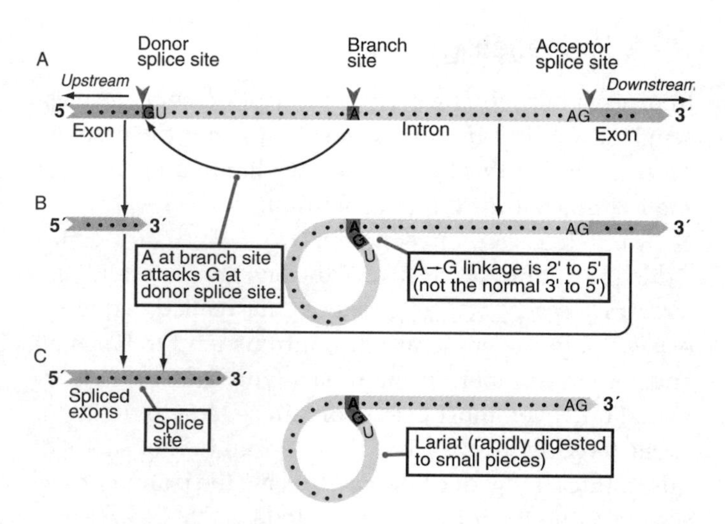

Figure 1-10 Removal of an intron from a primary transcript.

In fact, RNA splicing requires the participation of a complex set of small nuclear RNAs (snRNAs) and proteins that form a particle called a *spliceosome*.

The spliceosome is a dynamic entity; each of the snRNAs that participate in it also exist some of the time as individual RNA-protein complexes designated *snRNPs* (pronounced "snurps"). Assembly of the spliceosome begins with binding of U1 snRNP to the 5′ splice site, followed by binding of U2 to the branch point sequence. U6 and U4 (which is bound to U6) bind to U2, then U4 dissociates from U6 and U6 displaces U1. U5 brings the last nucleotide of the 5′ exon close to the first nucleotide of the 3′ exon, which facilitates joining of the two exons after the intron has been cut.

Alternative splicing is a major source of variability in gene expression, varying from one cell type to another and from one developmental stage to another. Recent surveys of expression of the human genome indicate that at least 40% of all genes undergo alternative splicing, and this may be an underestimate. In the simplest case, one exon may or may not be present in the mature mRNA. We do not yet have detailed knowledge of the mechanisms that lead to *exon skipping*, but it clearly is a common aspect of gene expression. Figure 1-11 There are also splicing variants that exclude only part of an exon and variants that include part of an intron. There are RNA-binding proteins that must bind to the splicing site before U1 binds and initiates spliceosome assembly. It is not yet clear how many such proteins exist. It is likely that the control of alternative splicing is a function of RNA-binding proteins (*splicing factors*) that affect commitment to splicing at specific introns. A group of RNA-binding proteins termed SR proteins (because they are rich in serine [S] and arginine [R]) are among the candidates being investigated. One member of that group, SC35, has been shown to have relatively high specificity for splicing of the beta-globin mRNA precursor. We can anticipate much more information on this form of RNA-processing control to be obtained in the next few years. Clearly, if gene-specific splicing factors exist, they may be an important source of mutants associated with genetic disease, as are variations in DNA sequence that affect the recognition of splicing signals by the processing machinery (Lecture 2).

Recently, it has been pointed out that, in some genes with many exons, there is the theoretical potential for production of thousands—even millions—of alternative transcripts from a single gene, but this is merely an exercise in *Gee Whizardry:* there is no evidence for that much variation. Nevertheless, it is certain that the transcriptome—the total variety of transcripts produced by a cell—exceeds the number of expressed genes.

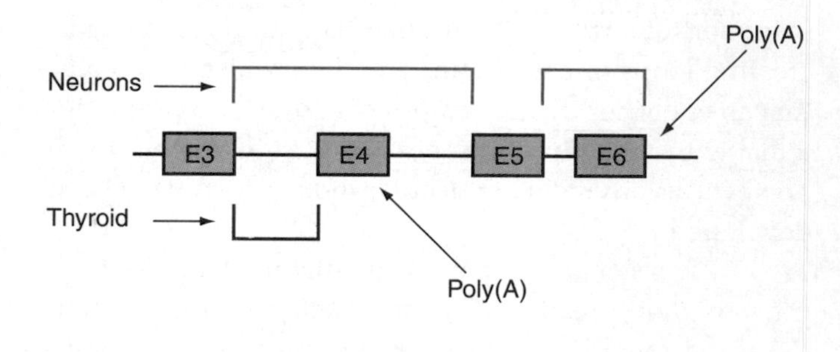

Figure 1-11 Production of distinct mRNAs by alternative splicing.

RNA processing also involves two types of terminal modification: capping at the 5′ end and polyadenylation at the 3′ end. Before the nascent transcript is more than a few dozen nucleotides long, the first nucleotide is "capped" by the addition of a *GMP* in a 5′ to 5′ linkage between riboses. The guanine is then *methylated* at the seven position, and sometimes another methyl group is added to the ribose of the next nucleotide. This cap is required for proper splicing of the first intron in pre-mRNA and for binding mRNA to ribosomes via a cap-binding protein. The cap also improves mRNA stability as well as transport of mRNA to the cytoplasm.

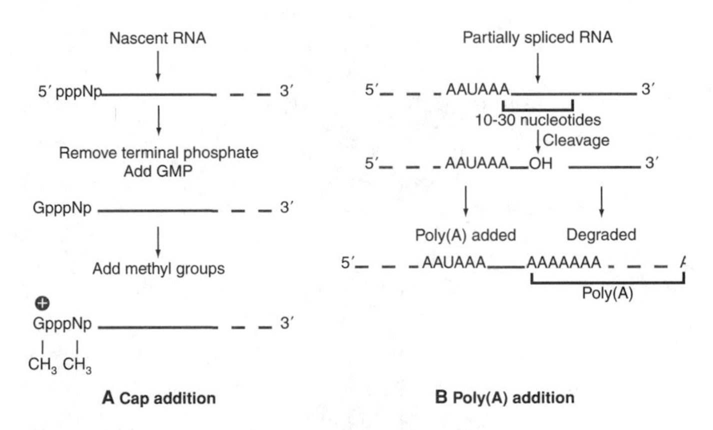

Figure 1-12 Terminal modifications of mRNA.

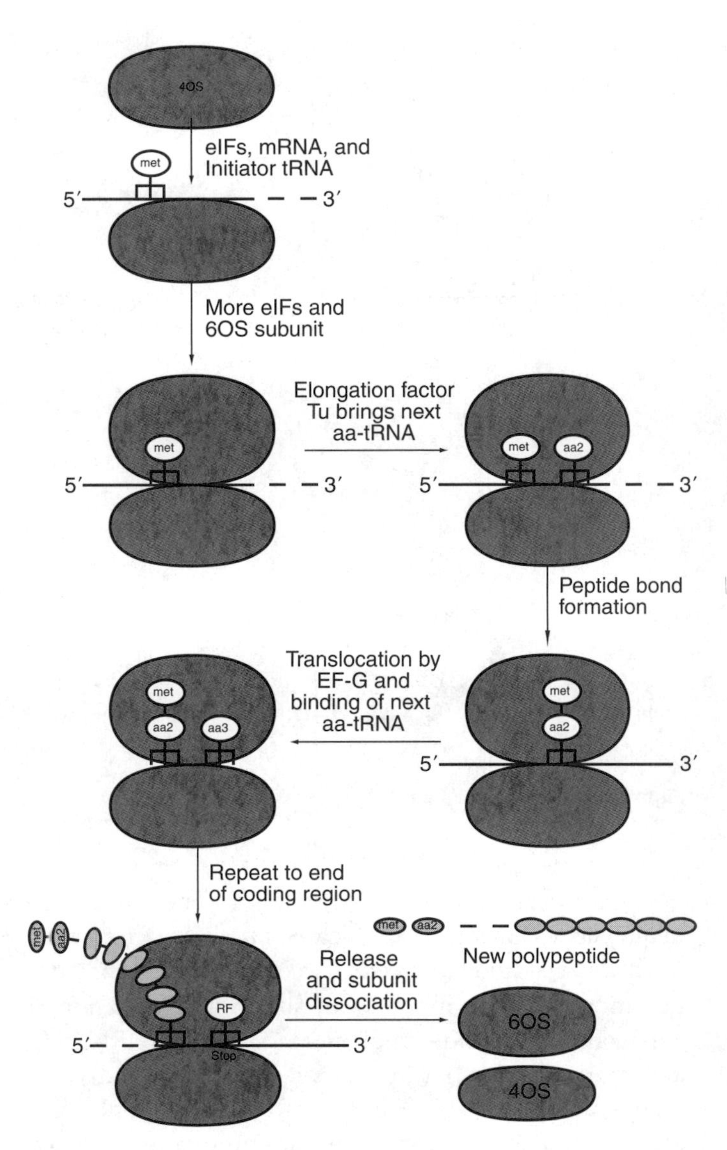

Figure 1-13 An outline of protein synthesis.

The 3′ end of the primary transcript is lengthened by *polyadenylation*, which is post-transcriptional addition of a series of adenosine nucleotides (usually 150 to 250), forming a *poly(A) tail*. The hexanucleotide *AAUAAA* occurs about 20 nucleotides upstream from the point at which poly(A) will be added; it is an important signal for *poly(A) polymerase*. During pre-mRNA processing in the nucleus, presence of a poly(A) tail is required for efficient removal of the last intron (but it does not affect removal of the other introns). In addition, *the poly(A) tail is an important regulator of mRNA stability*. It is gradually degraded by RNases in the cytoplasm and when it is gone, degradation of the mRNA quickly follows.

Translation

The word *translation* immediately suggests converting information expressed in one language into information expressed in a different language and that is exactly what happens in gene expression. Information is stored in the nucleic acid language in DNA as sequences of nucleotides; it is rewritten into the RNA dialect, then it is translated into the protein language using sequences of amino acids. A series of amino acids joined by peptide bonds can be called either a *polypeptide* or a *protein*. We usually refer to the translation of mRNAs as *protein synthesis*, although it certainly is polypeptide synthesis, also. When we want to emphasize that a functional protein contains the products of more than one gene, we may say that protein X consists of Y polypeptides (e.g., RNA and DNA polymerases are proteins that contain several polypeptides).

The major components involved in the translation process are mRNA, ribosomes, tRNAs, and a group of proteins known as translation factors. I have just described the production of mRNAs from primary RNA transcripts in eukaryotic cells. *Ribosomes* are often referred to as the factories where proteins are synthesized. They are complex macromolecular assemblages containing several RNAs and dozens of proteins. *Transfer RNAs (tRNAs)* are a group of about 30 small RNAs that have two essential properties. First, each type of tRNA can carry a specific amino acid at its 3′ end. The amino acids are attached to tRNAs by a group of enzymes called *amino acyl-tRNA transferases*. Second, each tRNA contains a set of three nucleotides called the *anticodon*, more or less in the middle of the tRNA, which are complementary to one of the *codons* that specify the corresponding amino acid in the genetic code (see Lecture 2, Table 2-1, for the full genetic code).

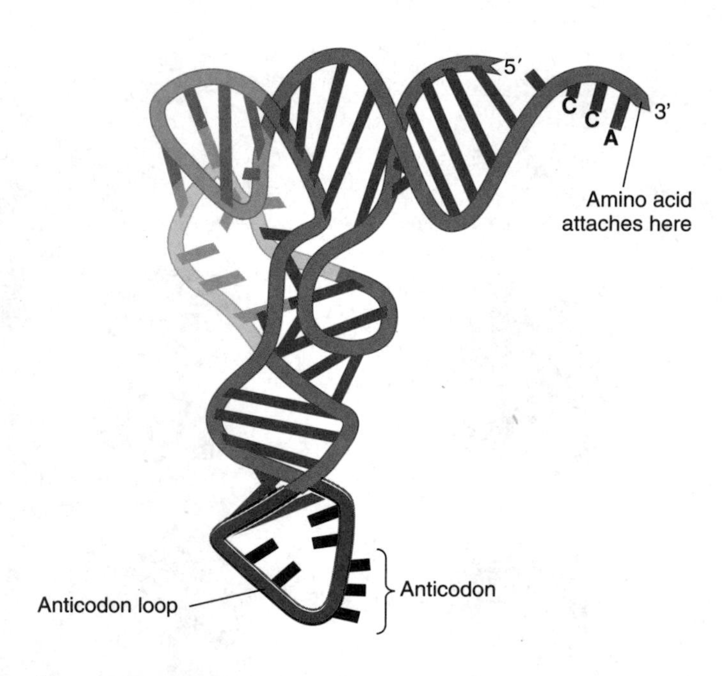

Figure 1-14 Structure of a tRNA.

The various *translation factors* are proteins that mediate each step in protein synthesis. There are three major classes of translation factors: *initiation factors, elongation factors,* and *termination factors.* We will consider initiation factors in more detail presently. Elongation of a nascent polypeptide involves binding of two tRNAs to a ribosome by base pairing of their anticodons with complementary triplets (codons) in mRNA. The amino acids carried by each tRNA are then brought together in such a way that they can be covalently joined by formation of a *peptide bond.* A fascinating fact is that catalysis of peptide bond formation is apparently done by one of the ribosomal RNAs, rather than by any of the dozens of ribosomal proteins.

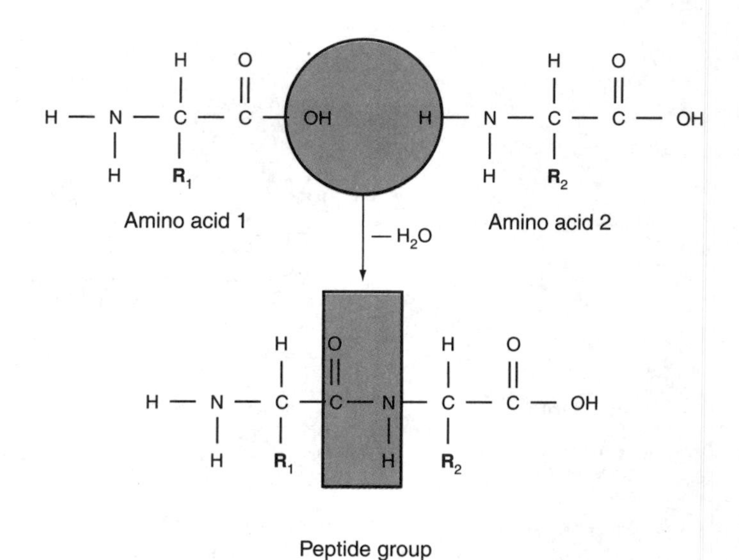

Figure 1-15 Peptide bond formation. R1 and R2 are the side chains that are unique to each amino acid.

Formation of a peptide bond also frees one of the tRNAs from its amino acid. That tRNA leaves the ribosome, the mRNA and the remaining tRNA move ahead one codon's distance on the ribosome, freeing up a binding site for the next tRNA, which must have an anticodon complementary to the next codon on the mRNA. Then the process is repeated over and over until a complete polypeptide has been synthesized.

Termination of protein synthesis in eukaryotes is carried out by two proteins (release factors): eRF1 recognizes all three stop codons and eRF3 is a ribosome-bound GTPase that provides the energy for release of the new polypeptide.

Protein synthesis is regulated in a complex manner. There are both general and specific aspects of translational control. General control points involve both ends of mRNA. One set of initiation factors binds to the 5′ methylated guanine, the "cap" referred to above, which permits binding of the small ribosomal subunit. Cap binding can be regulated by various other proteins. Another group of proteins binds to the poly(A) sequence at the 5′ end of mRNA, and in some cases, to sequences in the 5′ UTR near the poly(A). A surprising implication of recent research is that an actively translated mRNA is circularized, via protein–protein interactions at both ends of the mRNA. When the mRNA is not actively translated, the polyA becomes accessible to exonucleases, which shorten it and set the stage for degradation of the entire molecule.

The major control point for synthesis of specific proteins is initiation of the polypeptide chain, which always occurs at a methionine codon. Some mRNAs hide that initiator codon inside of loops or other structures formed by internal base pairing, so the participation of other macromolecules is necessary for translation to occur. In some cases, the regulatory molecule is a protein, but in other cases it is a small, noncoding RNA. Translation can also be regulated negatively, and there are cases where that is achieved because the polypeptide binds to its own mRNA, inhibiting further synthesis.

An example of a translational control region within the 5′ UTR is provided by the human *ferritin mRNA,* which contains a sequence of about 30 nucleotides that form a stem loop (a hairpin) via base pairing within the stem. This *iron response element (IRE)* can bind a protein that prevents translation initiation when the cell needs iron (and therefore does not need more ferritin, which is a protein that stores excess iron). When iron is abundant, the regulator protein binds iron and is released from the IRE on ferritin mRNA, allowing the synthesis of ferritin.

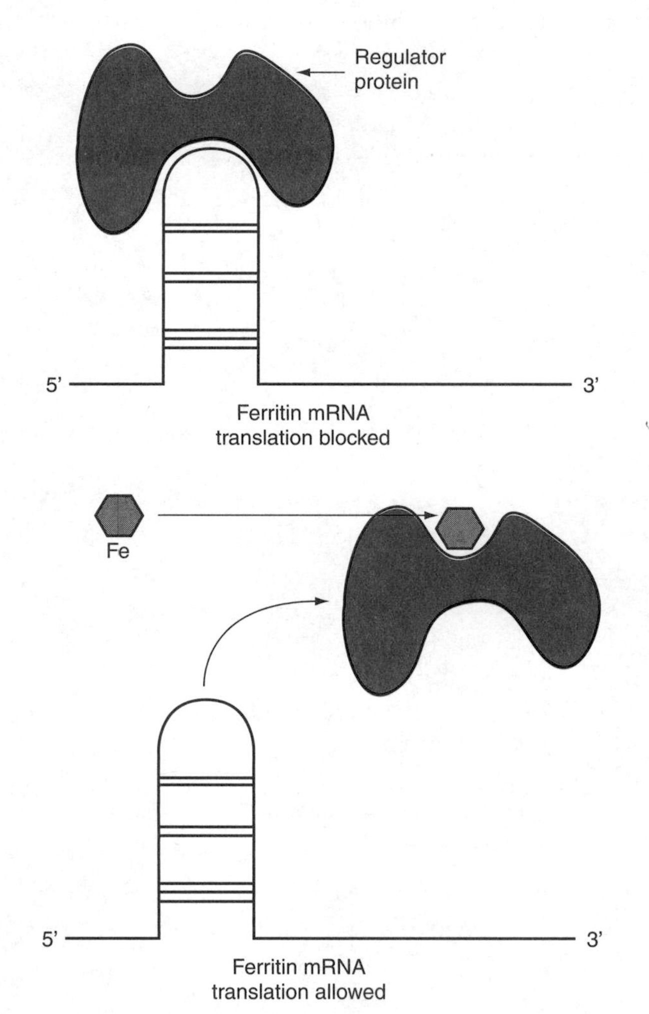

Figure 1-16 Control of ferritin synthesis.

Regulation of cellular iron content is also accomplished by changes in the stability of the mRNA for *transferrin receptor*. *Transferrin* is a protein that carries iron throughout the body; when it binds to the transferrin receptor on cell surfaces, iron can be taken into the cell. Thus, when iron is in short supply, cells need more transferrin receptor. In this case, the mRNA is inherently unstable. It contains five IREs in its long 3′ UTR, and when the regulatory protein binds to some of those IREs, it stabilizes the mRNA. If the iron supply rises, iron binds to the regulator, causing it to dissociate from the IREs. The transferrin mRNA then becomes susceptible to a specific endonuclease that cleaves it about 1 kb from the 3 end, and degradation occurs.

Protein phosphorylation also plays a role in translational control. One example occurs in *reticulocytes*, the precursors of red blood cells, which almost exclusively make hemoglobin. If the heme supply is limited, reticulocytes suppress the synthesis of alpha- and beta-globins via a protein kinase called *heme-controlled repressor (HCR)*. When HCR phosphorylates eIF2, it prevents binding of the initiator tRNA-Met to 40S ribosomal subunits, and protein synthesis cannot proceed.

The most recently discovered form of translational control is mediated by very small RNAs called *microRNAs (miRNAs)*, which average only 21 nucleotides in length. They originate as much longer transcripts that are processed into 60–70 nucleotide pieces that form stem-loop structures. In the cytoplasm, the loop and part of the stem are removed by an RNase called Dicer. In a manner yet to be elucidated, a single-stranded miRNA is generated, which forms a complex with some proteins; then the RNA portion presumably pairs with complementary sequences in the 3′ region of a specific mRNA and represses translation.

We do not yet know how common this form of translational control is in humans, but based on examples from the worm, *C. elegans*, it is likely to be important in development and differentiation (Lecture 7). MicroRNAs are similar to a class of molecules known as *small interfering RNAs (siRNAs)*, some of whose members function in destruction of viruses via pairing with a target RNA and cleavage of the target RNA by a specific RNase, which is part of the RNA-induced silencing complex (RISC). This process, known as *RNA interference (RNAi)*, is also being exploited with synthetic double-stranded RNAs to devise new ways of treating genetic diseases (Lecture 4). Note that *RNAi refers to a process*, not to a class of RNA.

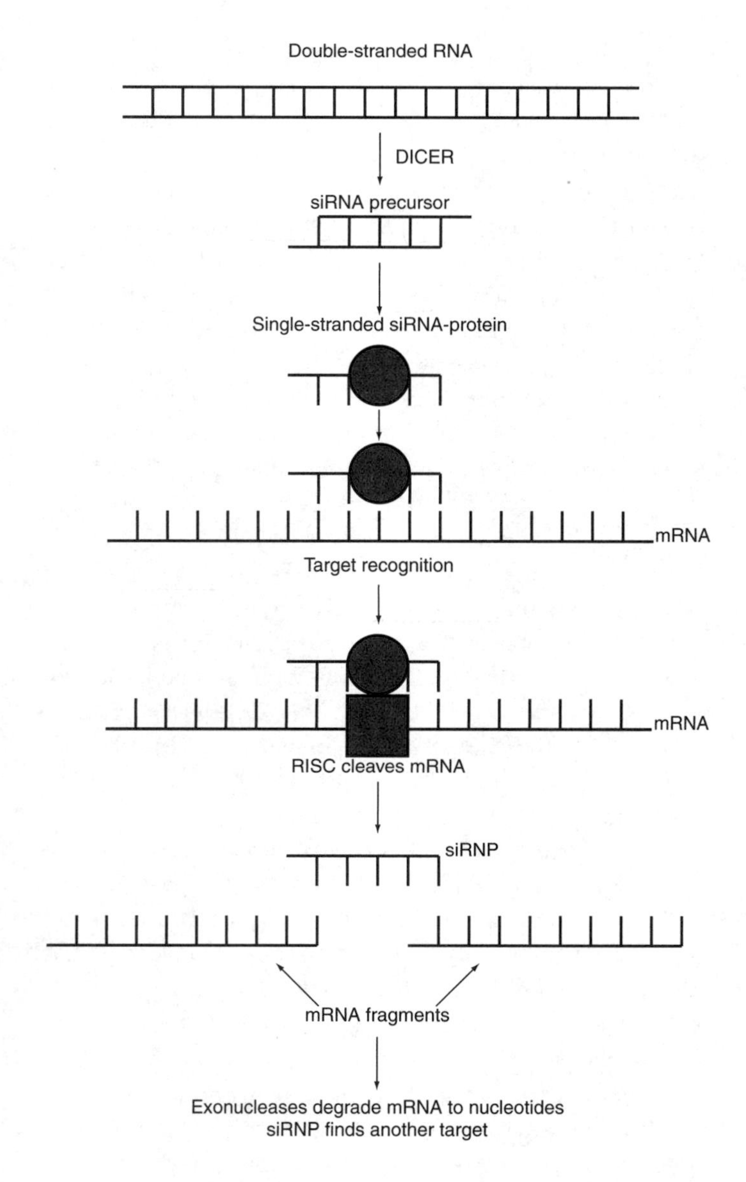

Figure 1-17 Translational control by siRNA.

These examples show that, as was the case for RNA synthesis, protein synthesis provides many opportunities for mutations in genes that code for regulatory molecules to affect the expression of a gene that is entirely normal. The complexities do not end there, because many proteins undergo post-translational modifications. These include the addition of various sugar molecules or fatty acids and the removal of small groups of amino acids (usually at the N-terminal end) in connection with sub cellular or extra-cellular localization of the protein. Another important class of post-translational modifications that we have already encountered is the addition of phosphate groups by *protein kinases* and the removal of phosphate groups by *protein phosphatases*, which often determine the catalytic activity of a protein and/or its interaction with other proteins.

The complexities of transcription, RNA processing, translation, and post-translational modifications lead to the conclusion that more than one polypeptide may be produced from one gene. Thus, the proteome—the total number of protein types produced by a given cell—significantly exceeds the number of genes expressed by that cell. Many laboratories are now engaged in developing techniques to analyze the proteome in ways that permit quantification of hundreds or thousands of polypeptides in a single assay. Proteome analysis will have an important impact on clinical practice (as well as on basic research) because it is the proteins that directly determine phenotype. Mutated genes and aberrantly processed RNAs do not by themselves make a person sick; morbidity is a consequence of the nonfunctional or malfunctional proteins that they encode. The major sources of mutations and their effects on gene expression will be described in the next lecture.

REVIEW

In this lecture we surveyed the general properties of the human genome: its size and organization into 23 pairs of chromosomes, the number of genes it contains, and the internal organization of genes. The virtually complete sequencing of the human genome by the Human Genome Project was summarized. You learned that gene duplication has created gene families, groups of genes that are related by descent, some of which have acquired new functions, while other have become inactive pseudogenes. The Human Genome Project has shown that almost half the human genome consists of reiterated sequences, which vary from a few copies to a million copies. The most abundant reiterated sequences (LINEs and SINEs) are retrotransposons or the descendants of retrotransposons, some of which can increase in number by reverse transcription of their own mRNAs and insertion of the corresponding cDNAs into new locations. Processed pseudogenes often arise by retrotransposition of mRNAs from other genes. All of these reiterated sequences are grist for the mill of evolution.

The association of DNA with a variety of proteins to form chromatin was described, with emphasis on the nucleosome, a complex of 200 DNA bp with an octamer of histones. Then we outlined the major steps in gene expression, which are (1) RNA synthesis, (2) processing of RNA to convert the initial transcript into mRNA, and (3) translation of the mRNA into a polypeptide. We emphasized that all aspects of gene expression require participation of many other molecules, in both basic processes and control of those processes. All of those functions can be affected by mutations in the genes for the proteins and RNAs involved, leading to a wide variety of inherited abnormalities.

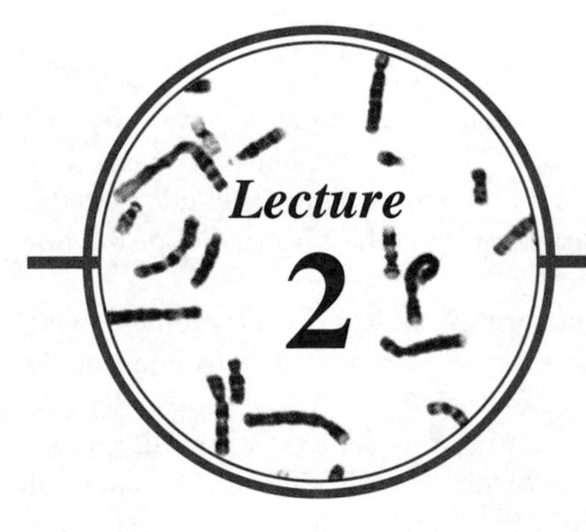

The Mechanisms of Mutation: How the Human Genome Changes

The human genome changes in many ways. Broadly speaking, a mutation is any permanent change in the DNA sequence. Mutations occur in germ cells and in somatic cells. Most mutations are benign, having no effect on the phenotype; however, when a mutation changes the coding sequence of a gene or the timing or extent of expression of a gene, a detectable abnormality may result. If such a mutation occurs in a germ line cell (one that gives rise to eggs or sperm) the abnormality may be manifested as a genetic disease in the next generation, but if the mutation occurs in a somatic cell, it may lead to cancer.

Mutations may be classified as spontaneous or induced. Spontaneous mutations are the result of normal DNA metabolism; they include spontaneous alterations of bases from thermal activity, errors in DNA replication, movement of transposable elements, and errors during meiotic recombination. Induced mutations arise when DNA is modified by agents that are not part of normal DNA metabolism. Those agents include radiation of various types, highly reactive metabolic byproducts that are produced within cells, and toxins that enter the body from direct contact, inhalation, or eating and drinking.

SPONTANEOUS MUTATIONS

As we enter this topic, it is important to understand that a change in the base sequence of DNA is not a mutation unless it becomes permanent; that is, it is passed on to daughter cells in a proliferating tissue or retained indefinitely in a post-mitotic tissue. For example, a common spontaneous change in DNA is *depurination*, removal of an adenine or guanine from a nucleotide as a result of random thermal fluctuations. It is estimated that up to 5,000 depurinations occur each day in an average human cell, and fewer than 1 in 1,000 depurinations fails to be repaired. Similarly, incorporation of incorrect nucleotides during DNA replication occurs routinely, but most of those mistakes do not become mutations. Cells possess complicated and generally effective DNA repair systems, some of which will be mentioned in several contexts later.

Mutations Arising During DNA Replication

Normal DNA replication is an extraordinary process. The 6 billion nucleotides in the nucleus of a human cell and the 46 chromosomes that contain them are so compacted and intertwined and bound with proteins that they make the Gordian Knot of legend look like simplicity. It is not surprising that DNA replication requires the activity of numerous proteins and that uncorrected mistakes are sometimes made.

The fundamental step in DNA replication is the creation of new strands of DNA complementary in sequence to the original strands. *DNA polymerases* are the enzymes that add nucleotides to each nascent DNA strand, using the bases on the template (old) strand as a guide, forming the standard Watson-Crick pairs of A with T and G with C. In eukaryotes, DNA polymerases can add no more than 100 nucleotides per second to a growing DNA strand. If they started at one end of a chromosome containing 100 million bp, it would take 1 million seconds (roughly 11 days) to get to the other end. However, the S phase of the cell cycle, during which the entire genome is replicated, usually takes only 6 to 8 hours. The problem is solved by having thousands of *origins of replication*; that is, sites where DNA synthesis can begin. Many replication origins are activated simultaneously, but not all of them. There is an orderly progression of DNA synthesis from one portion of the genome to another. In general, the heterochromatic portions of the genome are replicated late in the cycle.

Before polymerases can act, the DNA must be unwound and the two old strands must be locally separated. This is accomplished by several proteins, among which are *helicases* (which disrupt base pairs and open the double helix), *topoisomerases* (which break the sugar-phosphate backbone of DNA so that the stress caused by normal tight coiling does not prevent unwinding of the strands), and *single-strand binding proteins* (which help keep the two strands from re-pairing with each other before they have been replicated).

When a replication origin becomes accessible, synthesis of new DNA begins in a complicated manner. All known DNA polymerases, whether in mammals, plants, bacteria, or any other organism, are incapable of initiating a DNA strand. DNA polymerases require a *primer*, which is typically an RNA molecule 10–15 bases long. RNA polymerases do not require primers; one type of RNA polymerase acts as a *primase for DNA synthesis*; it binds to the origins of replication and synthesizes a primer there. Then DNA polymerase takes over, adding nucleotides one at a time in the 5′-to-3′ direction (that is, each nucleotide is joined to the existing strand by formation of a phosphodiester bond between the 3′ OH group on the ribose of the preceding nucleotide with the innermost phosphate [the alpha phosphate] of the nucleotide being added).

And that brings us to another problem with DNA replication: The two strands of DNA in a double helix are anti-parallel; that is, at any point the nucleotides in one strand are oriented in opposite directions. What is 5′-3′ on one strand is 3′-5′ on the other. All polymerases can only add nucleotides at the 3′ end of a strand, and yet a replication fork appears to grow in one direction along both template strands. One of the new strands can be made in a continuous manner, but the other strand has to be assembled discontinuously. What happens is that a primase makes RNA primers every few hundred nucleotides on the other template strand (called the *lagging strand*), then DNA polymerase fills in the gaps in the usual 5′-3′ direction and excises the primers when it comes to them. The fragments (called precursor or *Okazaki fragments*) are joined by *DNA ligase*. The net effect is that both nascent DNA strands grow in the same direction, although the strand being synthesized discontinuously actually lags the continuous strand somewhat.

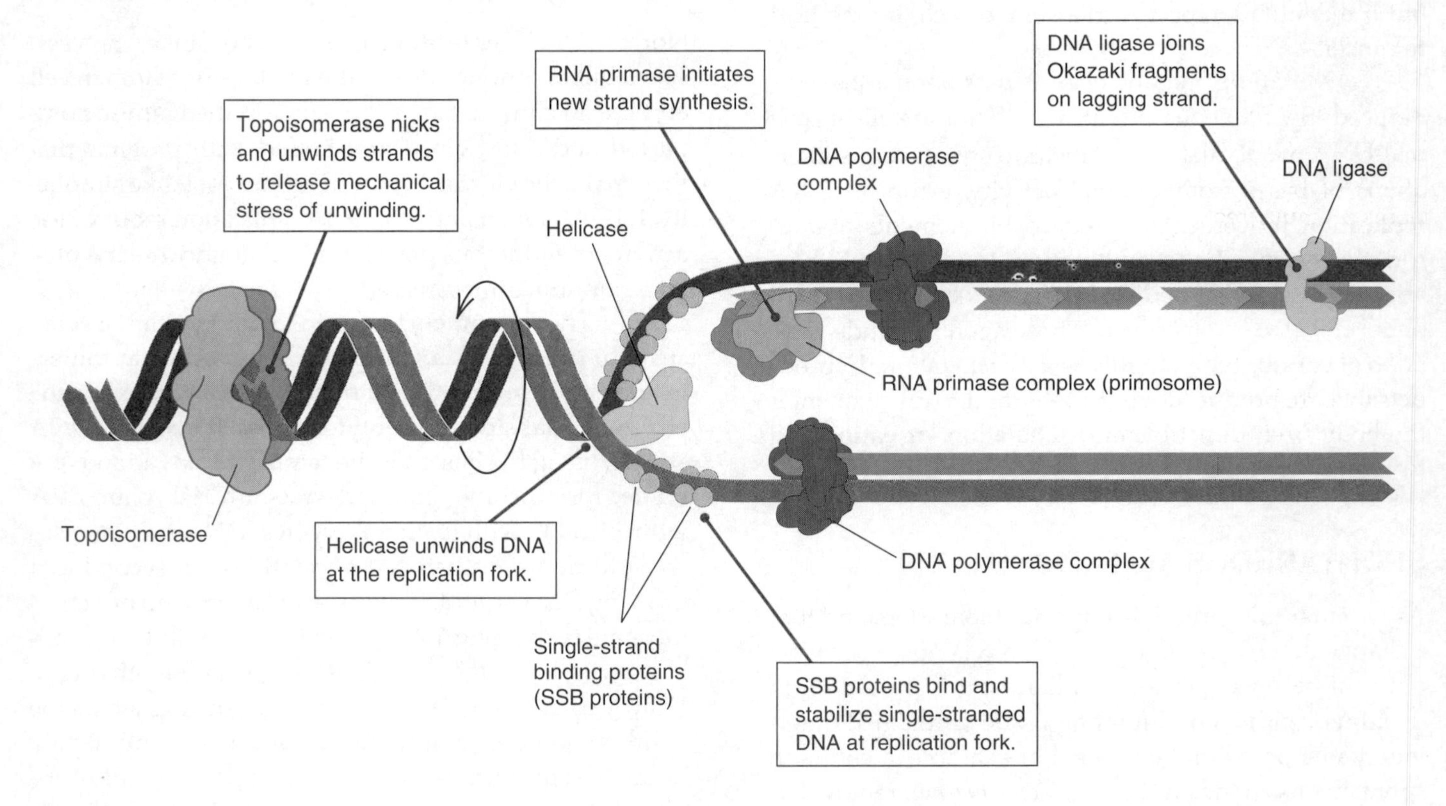

Figure 2-1 DNA replication.

This complex process of DNA replication proceeds with a high level of precision, but mistakes do occur. There are two general categories of mistakes: (1) incorrect incorporation and (2) replication slippage (strand slippage). Most of the mistakes in incorporation of nucleotides are immediately recognized as such and are corrected by one of several mechanisms. First, the principal DNA polymerases carry out a *proofreading* function; if the last nucleotide incorporated into a nascent DNA strand does not base-pair properly with the template strand, the polymerase has a 3′-5′ exonuclease activity that removes it. Other systems for *mismatch repair* exist that are able to correct incorporation mistakes that escape the polymerase's proofreading function. However, if an incorporation error is not corrected before the nascent strand is elongated substantially, it may remain and become a mutation. If an error persists until the next round of replication, half of the double-stranded daughter molecules will contain the original sequence and half will contain the mutation.

Mismatches in DNA cannot be blamed entirely on DNA polymerase. Some are the result of a low spontaneous rate of change in some of the bases. The most common change from one base to another is deamination of cytosine, which creates a uracil residue. Uracil would pair with adenine in the next round of replication, thus converting a GC pair to an AT pair. Fortunately, there is an enzyme that recognizes uracil and removes it; another enzyme then removes the exposed deoxyribose-monophosphate, and the resulting single-stranded gap is repaired by one type of DNA polymerase, thus restoring the original sequence. No problem there. However, cytosine is frequently methylated (mentioned in Lecture 1) and when 5-methyl cytosine deaminates, it becomes thymine. Obviously, DNA polymerase cannot tell a spurious thymine from a legitimate thymine, so the polymerase places an A opposite the new thymine when it replicates that strand of DNA, and a mutation is established. Numerous other base modifications can be caused by alkylating agents that enter the body with food or by inhalation, and ultraviolet light can produce pyrimidine dimers in exposed areas, as will be described later in this lecture. All such changes may cause mutations when the damaged DNA is replicated. There will be more about DNA repair mechanisms in Lecture 6.

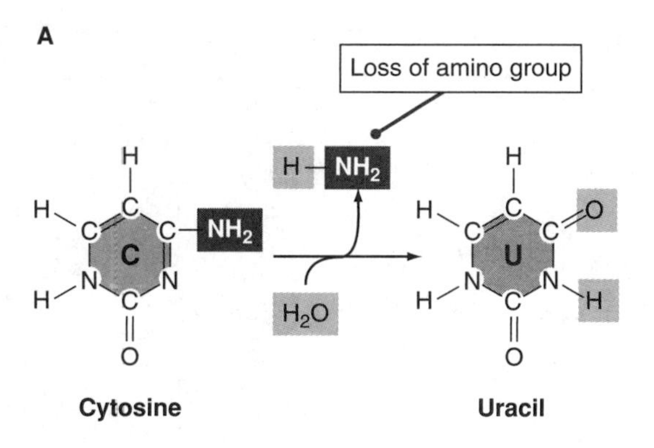

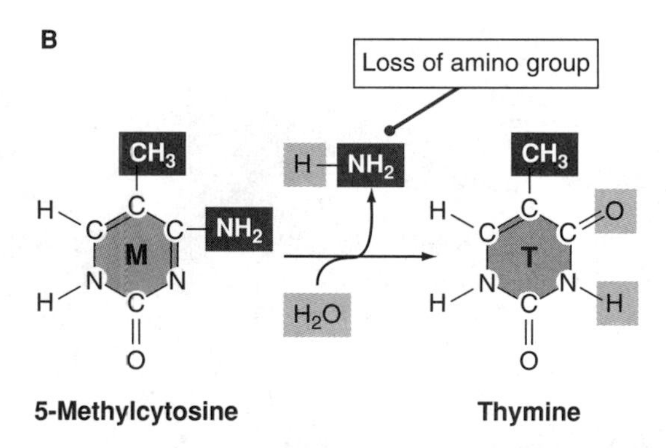

Figure 2-2 Spontaneous changes of cytosine to uracil and 5-methylcytosine to thymine.

Replication slippage is believed to be a major mechanism of mutation, especially in areas where there are tandemly repeated sequences. To visualize this, you should realize that the molecules being replicated are not passive; they are continually twisting and gyrating and bouncing off of other macromolecules. Understandably, this leads to frequent separation of the last few newly added nucleotides from the template strand, and when that loose end re-forms base pairs with the template DNA, it may not pair with its legitimate target. This is easiest to see when a *short tandem repeat (STR)* sequence is being replicated. Whenever a few base pairs are transiently broken, the two strands may not line up again perfectly, because normal base pairs can be established between any of the repeat units.

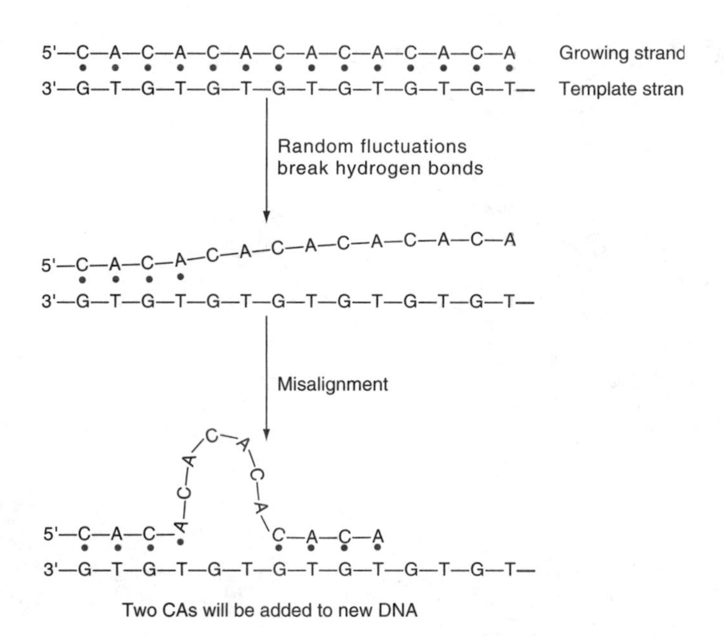

Figure 2-3 A short tandem repeat (CACACA . . .) sequence, with a looped-out segment and imprecise re-pairing.

STRs have an extremely high mutation rate, with some of them changing the number of repeats at a frequency of 1% per generation in humans. Increases in repeat frequency are more common than decreases, but both types of change occur often. However, replication slippage does not *require* tandem repeats; it may take place anywhere that a transient mispairing between the end of the nascent strand and the template strand can form and persist long enough for the polymerase to be deceived into continuing its work from the wrong point on the template.

There are thousands of STRs in the human genome, and their high mutation rate means that there are frequent STR *polymorphisms* in the population. That is, individuals often have different sized alleles at a given locus, with each allele consisting of different numbers of the basic repeat unit. These polymorphisms make it possible to develop a *DNA fingerprint* for every individual that is unique to that individual (except for identical twins). The ability to identify every person via their DNA fingerprints has been the basis for important developments in forensic genetics. STR polymorphisms have also been used to trace the history and movements of human populations and in linkage studies to help identify genes that underlie a particular disease.

One form of tandem repeats is a significant source of human genetic disease. Some genes contain sequences in which the same codon is repeated several times in an unbroken series. These *trinucleotide repeats* (or *triplet repeats*) are subject to replication slippage (as diagrammed in Figure 2-3), and it seems that the more of them there are, the more likely it is that replication slippage will occur, usually causing the repeat series to grow even longer. When that gene is expressed, an abnormal protein may be produced, which may lose its function or interfere with the function of other proteins in some way. A well-known example of *triplet repeat disease* is Huntington disease (where normal persons have 10–35 copies of a CAG repeat, which encodes a polyglutamine segment of the protein; affected persons have from 36 to more than 120 copies of the repeat—see Lecture 4). Some other triplet repeat mutations are in noncoding regions of a gene. An example is myotonic dystrophy, where a CTG repeat in the 3′ UTR of the messenger RNA for a protein kinase may be expanded to as many as 3,000 copies, which interferes with processing the mRNA.

Mistakes in DNA replication are the presumed basis for the fact that the mutation rate is higher in male mammals than in females. Studies on mice first indicated that the mutation rate is five to six times higher in male gametes than in female gametes. The explanation is simple: A mature *primary oocyte* in humans has gone through approximately 24 cell divisions since the formation of the zygote, whereas in a man 25 years old, a *sperm* is the product of about 300 cell divisions. Oocytes are all formed before a human female is born, but sperm are produced continuously throughout most of adult life. The older the father, the more opportunity his sperm have had for mistakes in DNA replication, and this is confirmed by studies that show a higher frequency of genetic diseases among the children of older males.

Effects of Changes in Base Sequence on Gene Expression

The most frequent class of replication errors leads to *single nucleotide substitutions*. If a substitution takes place in a coding sequence, it may or may not produce a change in amino acid sequence of the encoded protein. The genetic code is *degenerate*, which means that most amino acids are encoded by more than one trinucleotide (see Table 2-1). This is especially true for the third position of most codons, where substitutions of one nucleotide by another usually do not change the amino acid. For example, the codons GGU, GGC, GGA and GGG all specify glycine. Such substitutions are called *synonymous mutations*; mutations that do cause an amino acid substitution are called *nonsynonymous or missense mutations*.

Table 2-1 The standard genetic code

First position (5′ end)	**Second position** U	C	A	G	**Third position (3′ end)**
U	UUU Phe UUC Phe } **F** UUA Leu UUG Leu } **L**	UCU Ser UCC Ser UCA Ser UCG Ser } **S**	UAU Tyr UAC Tyr } **Y** UAA Stop UAG Stop	UGU Cys UGC Cys } **C** UGA Stop UGG Trp **W**	U C A G
C	CCU Leu CUC Leu CUA Leu CUG Leu } **L**	CCU Pro CCC Pro CCA Pro CCG Pro } **P**	CAU His CAC His } **H** CAA Gln CAG Gln } **Q**	CGU Arg CGC Arg CGA Arg CGG Arg } **R**	U C A G
A	AUU Ile AUC Ile AUA Ile } **I** AUG Met **M**	ACU Thr ACC Thr ACA Thr ACG Thr } **T**	AAU Asn AAC Asn } **N** AAA Lys AAG Lys } **K**	AGU Ser AGC Ser } **S** AGA Arg AGG Arg } **R**	U C A G
G	GUU Val GUC Val GUA Val GUG Val } **V**	GCU Ala GCC Ala GCA Ala GCG Ala } **A**	GAU Asp GAC Asp } **D** GAA Glu GAG Glu } **E**	GGU Gly GGC Gly GGA Gly GGG Gly } **G**	U C A G

Note: Each amino acid is given its conventional abbreviation in both the single-letter and three-letter format. The codon AUG, which codes for methionine (boxed) is usually used for initiation. The codons are conventionally written with the 5′ base on the left and the 3′ base on the right.

Changing one amino acid in a protein may or may not have functional consequences. If the change affects the catalytic site of an enzyme or drastically alters the configuration of the protein, the protein's function may be abolished. A variant gene that cannot produce any functional product is called a *null allele.* The opposite alternative is that no detectable change in the protein's function may result, or even that an increase in the catalytic activity of an enzyme may occur. Between those extremes there is a spectrum of possibilities for partial function, some of which may affect the catalytic activity of an enzyme and some of which may alter the interactions of a protein with other macromolecules. The well-known sickle-cell mutation in the sixth codon of the beta-globin gene changes GAG (glutamine) to GTG (valine), which affects the interaction of globin polypeptides with each other, causing aggregation that distorts the shape of red blood cells. A succinct way to annotate simple mutations uses the one-letter amino acid code (see Table 2-1); the sickle cell mutation is Q6V.

Nonsense mutations arise when a nucleotide substitution converts a coding triplet to one of the three *stop codons* (termination signals for protein synthesis). The name refers to the fact that the codon no longer specifies an amino acid. Single nucleotide substitutions can produce stop codons in a variety of ways. The diagram shows some examples. Nonsense mutations often are null alleles, because all of the amino acids downstream from the mutation are missing. However, it is possible to have a nonsense mutation near the 3′ end of a gene (which corresponds to the carboxyl terminus of a protein) that does not delete enough information to have a significant phenotypic effect.

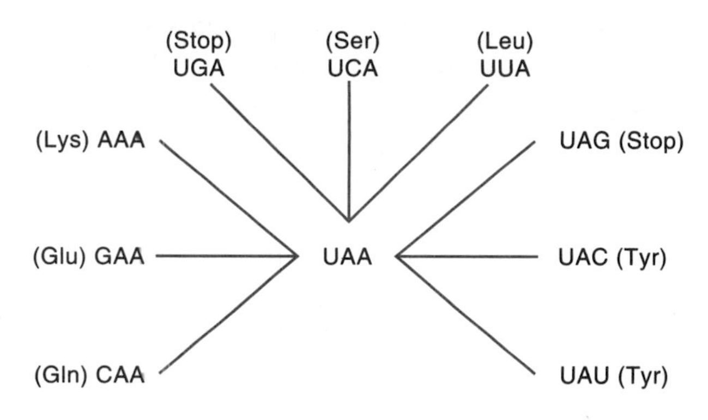

Figure 2-4 Nine ways to produce a stop codon from single nucleotide substitutions in other codons.

Anti-termination mutations arise when a mutation changes a stop codon to one encoding an amino

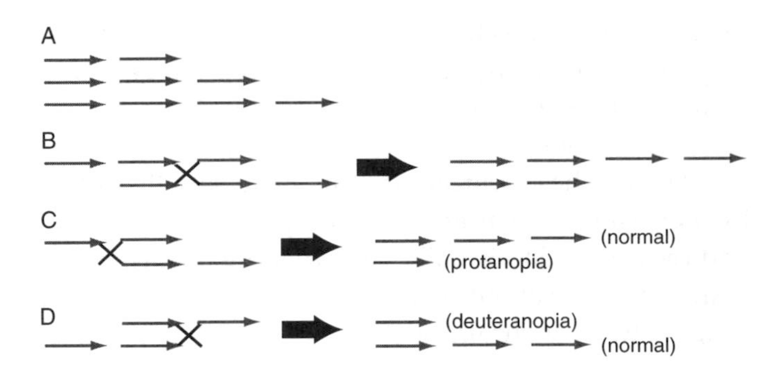

Figure 2-9 Normal X chromosomes have one red pigment gene (black arrow) and one, two or three green pigment genes (green arrows, panel A). Unequal crossing-over can alter the number of green pigment genes (panel B), or delete the red pigment gene (panel C), or delete the green pigment genes (panel D).

Gain or loss of DNA segments can also occur during the somatic cell cycle, where a process known as *sister chromatid exchange (SCE)* often occurs after DNA synthesis but before mitosis (i.e., in G2 phase). Because the DNA sequences of sister chromatids are identical, SCE normally leads to no changes. However, the presence of dispersed repeated DNA elements like LINEs and SINEs opens many possibilities for unequal sister chromatid exchange, leading to duplications and deletions that will be reproduced in subsequent cell divisions. If unequal SCE takes place in a germ cell precursor, mutation-bearing gametes may result. *Mitotic recombination* (exchange of DNA between chromatids on the homologous members of a chromosome pair) also occurs, although it is relatively rare. Like meiotic recombination or SCE, it is subject to errors.

Chromosomal Mutations

Changes in the structure of chromosomes that are large enough to be detectable microscopically are termed chromosomal mutations or chromosomal rearrangements. Most rearrangements are either inversions, translocations, or deletions, although there are a few rarer categories.

Inversions are rearrangements that alter the polarity of a chromosomal segment, relative to the rest of the chromosome. Inversions can result from a chromosome being broken in two places, then resealed with the broken segment in the opposite orientation (although there are other mechanisms). If the breaks do not have any adverse effects on structure or expression of essential genes, organisms with an inverted chromosome are viable. However, problems can arise at meiosis in an individual heterozygous for an inversion. Synapsis at meiotic prophase aligns homologous DNA sequences on a gene-by-gene basis. Therefore, for a chromosome with the normal sequence to pair with a chromosome with an inversion, one of the two must form a loop, as shown in Figure 2-10. In many cases this is easily done and has no deleterious consequences, unless crossing-over takes place within the loop, between one normal and one inverted chromatid. If that occurs, each daughter chromatid will be duplicated for some genes and deleted for some other genes. When those chromatids are segregated into germ cells and participate in zygote formation, an inviable embryo will usually be formed. The two chromatids that did not exchange DNA will have the usual set of genes (one in normal order and one in inverted order) and will be able to participate in normal gametes. Thus, a heterozygous inversion will lead to a reduction in fertility but not to sterility.

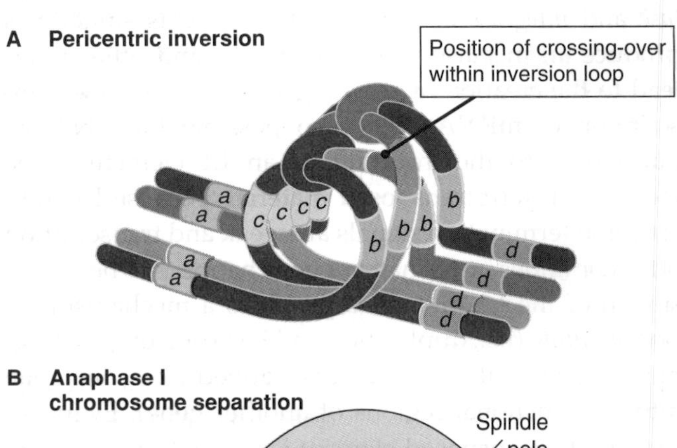

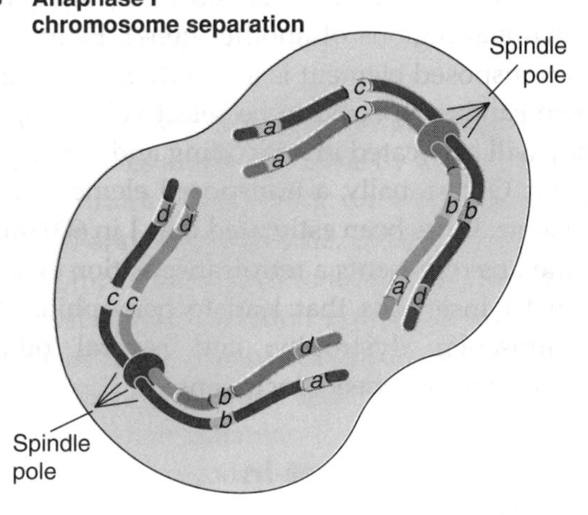

Figure 2-10 Effect of crossing-over when one member of a homologous chromosome pair has a pericentric inversion. (A) Configuration at synapsis. (B) Configuration at anaphase: two of four chromatids are abnormal.

Translocations are exchanges of genetic material between nonhomologous chromosomes. Often, the translocations are *reciprocal*, with no loss of DNA from either chromosome. Nevertheless, the translocations may affect gene structure or expression; some examples that are involved with cancer cells will be given in Lecture 6. Other times, there is no apparent phenotypic effect. Many individuals with benign reciprocal translocations have been identified as a byproduct of karyotyping done on women who are at elevated risk for a Down syndrome fetus, because of their age (Lecture 4).

But once again, trouble arises during meiosis. The two translocated chromosomes and their normal homologues try to pair gene-by-gene in meiotic prophase, forming a cross-shaped structure (see Figure 2-11). At anaphase, that structure can be pulled apart in three different ways, two of which lead to gametes with partially abnormal genomes. Sometimes persons who are reciprocal translocation carriers (males or females) are identified because a woman has several miscarriages. An inversion can have the same effect.

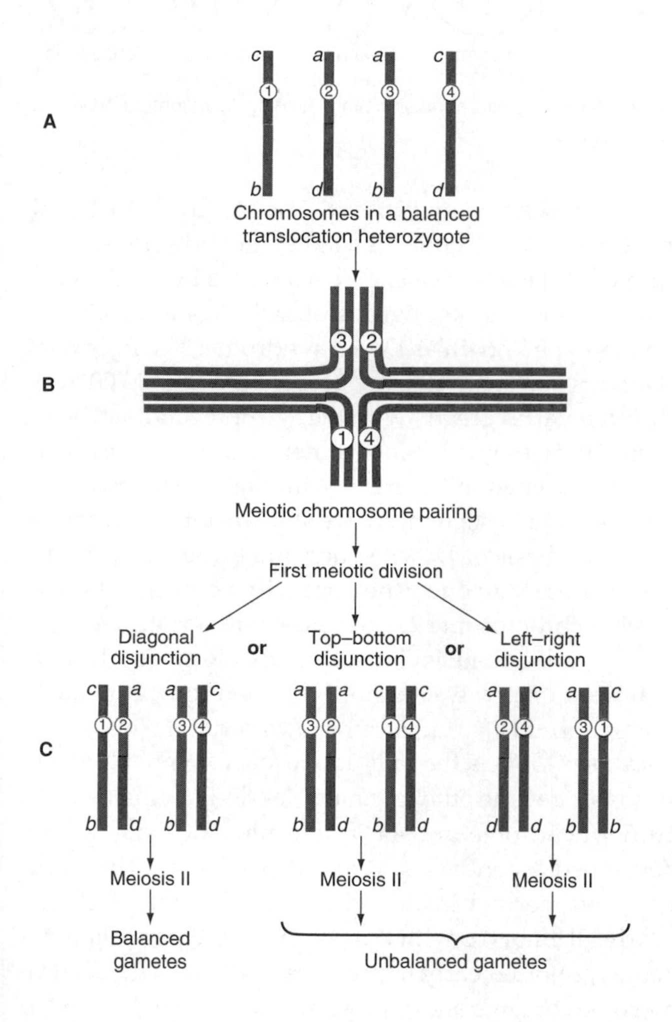

Figure 2-11 Meiosis in a heterozygote for a reciprocal translaocation. (A) Pairing between two normal and two translocated chromosomes. The chromosomes can be pulled apart at anaphase I in three ways: top-bottom, left-right, or diagonally. (B) Chromosomes in gametes that result from the three ways of separating chromosomes in A.

A relatively common form of nonreciprocal translocation involves joining the long arms of two acrocentric chromosomes at their centromeres, with loss of the short arms. These are called *Robertsonian translocations*. Because the short arms of acrocentric chromosomes apparently contain only redundant sequences, such as genes for rRNA, the loss of two short arms out of a total of 10 (see Lecture 1) is not noticeably deleterious. However, at meiosis the same problem that was described for reciprocal translocations arises, and many of the gametes produced by a carrier of a Robertsonian translocation are abnormal, either having only one copy or having three copies of a portion of one chromosome. Chromosome 21 is often involved in Robertsonian translocations, which are the source of a fraction of children born with Down syndrome.

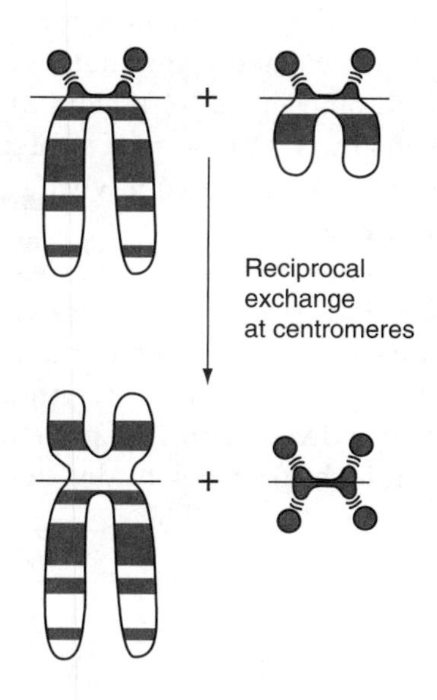

Figure 2-12 A Robertsonian translocation.

Deletions of any size may occur in chromosomes. They may be *terminal* (loss of material from one end of a chromosome) or *interstitial* (internal loss). Deletions do not lead to meiotic difficulties, but depending on the genes lost, they may have serious phenotypic consequences. One relatively common (1 in 50,000 births) terminal deletion in humans involves the tip of chromosome 5p. It causes Cri-du-Chat syndrome (cat-cry), so-called because newborns make a sound like a kitten. Although affected persons may survive many years, they are severely retarded mentally and have various physical abnormalities.

Abnormal numbers of one or more chromosomes are a common type of chromosomal mutation, especially in spontaneous abortions. Geneticists use a form of shorthand for describing the number of chromosomes in an individual; thus, the normal human genome is (46,XX) for females and (46,XY) for males. *Polyploidy* is the presence of extra complete sets of chromosomes. *Triploid babies* (three sets of chromosomes, 69,XX or 69,XY) are occasionally born alive, but they do not survive more than a few days. Triploid embryos are much more common, accounting for about 10% of spontaneous abortions in the first trimester of pregnancy.

Aneuploidy is the occurrence of an abnormal number of a specific chromosome (i.e., any number other than two copies for autosomes or anything other than two X chromosomes in females or one X and one Y in males). Aneuploidy is much more common than polyploidy. *Most autosomal aneuploidies are trisomies*; that is, the presence of three members of one chromosome type in otherwise diploid cells. Apparently only three autosomes can be trisomic and compatible with live births: They involve chromosomes 13, 18, and 21. *Monosomies*, the presence of only one member of a chromosome pair, do not occur in live births, except when one X chromosome is missing in females.

Aneuploidies are usually the result of nondisjunction at one of the meiotic divisions, which is the failure of chromosomes to be distributed normally to daughter cells. Figure 2-13 illustrates nondisjunction at the second meiotic division during spermatogenesis, but it can also occur in the first meiotic division in either sex.

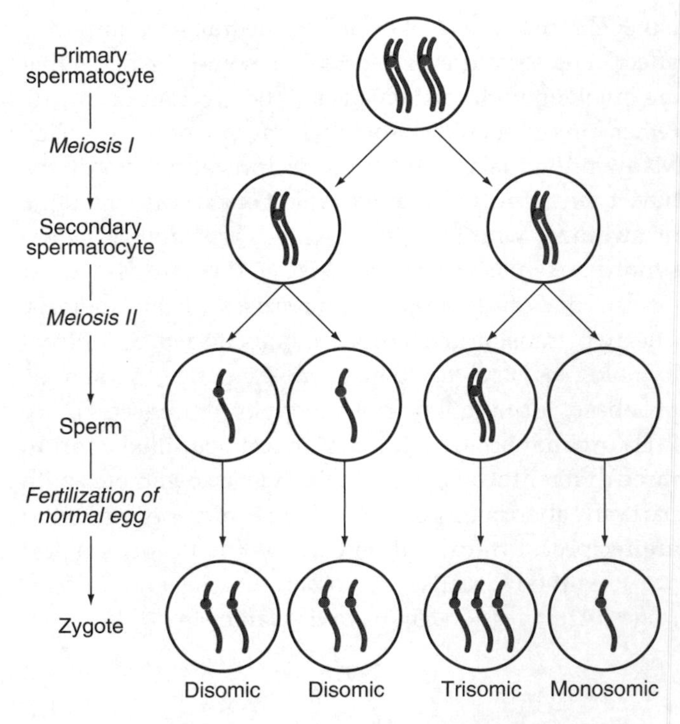

Figure 2-13 Meiotic nondisjunction, leading to trisomic and monosomic zygotes.

Trisomy 13 and *trisomy 18* occur in about 1 birth in 5,000–15,000. Neither is viable; most do not survive for more than a few weeks. *Trisomy 21* is by far the most common and best known autosomal trisomy, producing a phenotype known as Down syndrome. The frequency of infants born with trisomy 21 is roughly 1 per 700 overall, but it varies greatly with the age of the mother, being about 50 times greater for women who are 45 years old than for women in their 20s. Although chromosome 21 has been totally sequenced, we still are not sure what the molecular basis of Down syndrome is (but it clearly involves more than one gene), nor why nondisjunction involving chromosome 21 increases with oocyte age.

Aneuploidies involving the X or Y chromosomes are relatively common, are often compatible with nearly normal life, and occur in a variety of forms. *Turner syndrome (45,X)* is the only monosomy compatible with survival beyond the neonatal period. Females with Turner syndrome are short, infertile, and have poorly developed secondary sexual characteristics. They usually have normal intelligence. It is a curious fact that nearly all embryos with the (45,X) karyotype are spontaneously aborted early in pregnancy. Perhaps the X chromosome contains one or more genes that usually need to be expressed from two copies at some stage of early development, but why some embryos develop to birth with only one copy is not known.

Another common aneuploidy is the presence of an extra X chromosome in persons who are phenotypically male, which is called *Klinefelter syndrome (47,XXY)*. Persons with this karyotype are sterile and have ambiguous secondary sex characteristics, although they tend to be tall. Intelligence varies from normal to mildly retarded. Males with an extra Y chromosome (47,XYY) are about as common as Klinefelter males (1 in 1,000 births). They are usually tall and are fertile. They have a tendency to be impulsive and aggressive, which sometimes gets them in trouble with the law. However, the efforts by journalists some years ago to conclude that the Y chromosome contains a "criminal gene" were just speculation. Most men with an extra Y chromosome lead normal lives and never know that they have an unusual karyotype.

INDUCED MUTATIONS

In addition to the various mechanisms described in the previous sections, mutations may be produced by ionizing radiation, ultraviolet radiation, and chemical mutagens.

The first source of induced mutations to be discovered was X-rays. Hermann Muller demonstrated in 1927 that X-irradiation of *Drosophila* produced an increase in the frequency of several easily observed mutant phenotypes. Moreover, the increased rate of mutant production was proportional to the dose of X-rays administered. X-rays are a type of *ionizing radiation*, which also includes gamma rays, as well as the alpha and beta particles emitted by radioisotopes. They all produce *free radicals*, which are highly reactive molecules with unpaired electrons. Free radicals are the vandals of the intracellular world. They quickly and indiscriminately modify other molecules. Free radicals can break DNA strands and can also modify DNA bases in several ways. X-rays generate so many free radicals that double-strand breaks in chromosomes are a frequent consequence. Although cells have mechanisms that can repair double-strand breaks, repair is not always complete, and cells with chromosomal fragments that lack a centromere are likely to lose essential genetic elements the next time they divide. Alternatively, repair of double-strand breaks may produce *dicentric* chromosomes (with two centromeres) or other abnormal structures that also lead to loss of genetic material at mitosis.

Radioisotopes also produce ionizing radiation. The children of people exposed to ionizing radiation by the atomic bombs that were dropped on Japan in 1945 were the subjects of an intensive study for the next several decades. It will be described in more detail in "The Human Mutation Rate" section. More recently, the explosion of a nuclear reactor in Chernobyl, Ukraine, on April 26, 1986, provided another unplanned human experiment. As the radiation cloud drifted north, many people in the neighboring republic of Belarus were heavily exposed to radioisotopes, principally I-131 and Cs-137. The full extent of the genetic damage they suffered is not yet known, but one study has clearly shown an increase in the mutation rate at several short tandem repeat loci.

Ultraviolet radiation (UV) can produce several types of alterations in DNA bases, the most common being *pyrimidine dimers* (thymine or cytosine). The dimers are formed between adjacent pyrimidines on one strand of DNA and are potentially capable of blocking both RNA and DNA synthesis, because the bases in the dimer cannot participate in normal base pairing. For humans, the principal source of UV exposure is sunlight, and both the medical and popular literature are full of admonitions to limit one's exposure. The warning is particularly applicable to persons who ski or hike at high altitudes and to persons at high latitudes in the southern hemisphere, because of the "ozone hole" (thinning of the ozone layer over Antarctica, which would normally absorb a large portion of UV light).

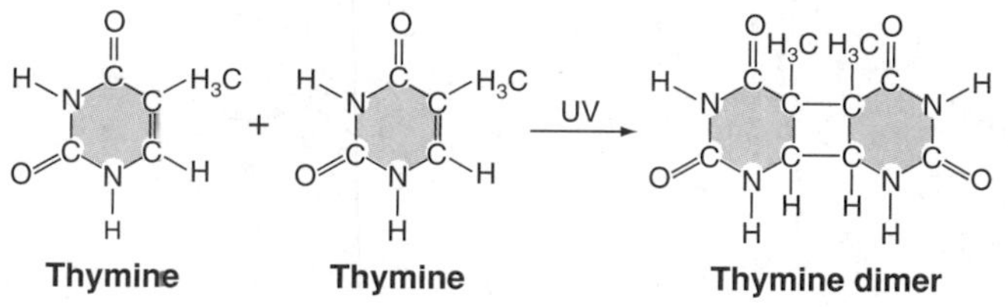

Thymine + Thymine —UV→ Thymine dimer

Figure 2-14 Diagram of a thymine dimer.

Most organisms have an enzyme that absorbs light and uses that energy to cleave pyrimidine dimers, a process known as *photoreactivation*, but placental mammals, including humans, do not. We have DNA polymerase *eta*, which insert dAMPs across from pyrimidine dimers. If the dimer was made of thymines, that repairs the defect. If it was made of cytosines or a cytosine and a thymine, a mutation is created. Various other effects of UV irradiation can be corrected by several other enzymes. Persons with a defect in any of the corresponding genes suffer from a condition called *xeroderma pigmentosum*, which will be described in Lecture 6.

Free radical production is also an unfortunate consequence of normal metabolism. In Lecture 5 you can read about energy production in mitochondria, which depends on the transfer of electrons through a series of carrier molecules, ending with molecular oxygen (O_2). The process is not perfectly efficient, and one nasty byproduct is the superoxide radical (O_2^-). Superoxide and a variety of related compounds are believed to be responsible for a significant portion of the high mutation rate in mitochondrial DNA and the accumulation of defects in mitochondrial function that may be an inescapable aspect of aging. Free radicals are also generated within lysosomes of macrophages and neutrophils as part of the process of degrading ingested bacteria and other substances.

We are also at risk from *chemical mutagens*. Partly because of naturally occurring toxic compounds and partly as a consequence of our technological prowess, we live in a dangerous world. Some of the chemicals produced for industrial purposes are known mutagens and thousands more are potential mutagens. They are present in the air we breathe, the fluids we drink, the foods we eat, the clothes we wear, and virtually everywhere else. Studies on experimental organisms have allowed us to learn a lot about how chemical mutagens act. The subject is too complex to review thoroughly here, so we will mention only a few categories.

Some complex organic dyes (such as acridine or ethidium bromide) look rather like a *purine-pyrimidine pair*. They can *intercalate* between the bases in DNA, thus stretching the double helix. When the DNA is replicated, a base pair may be added or deleted; this produces a *frameshift mutation*, which usually has serious consequences if it occurs within a coding sequence, as described earlier in this lecture.

Another large class of organic molecules act as *alkylating agents*, modifying the bases of DNA at one of several places. One such compound used to induce mutations in the laboratory is *ethyl methane sulfonate*, and one of its effects is to place an ethyl group on guanine or thymine. Ethylated guanine can form base pairs with thymine, which then leads to a GC pair being replaced by an AT pair. Ethylated thymine can form base pairs with guanine, which leads to an AT pair being replaced with a GC pair.

The foods we eat are a major source of chemical mutagens. Did you know that the fresh mushrooms you had on your salad contain hydrazine? That highly reactive compound and its derivatives can modify DNA in several ways, including breakage of strands. Another important class of potential mutagens is heterocyclic amines, which are generated when meats are cooked well-done. Derivatives of heterocyclic amines form addition products on DNA bases, leading to mutations when DNA is repaired or replicated.

A frustrating fact is that our bodies contain a complex system of detoxifying enzymes that sometimes make a bad situation worse. The *P450 enzymes*, most of which function in the liver, are the product of more than 50 genes. Their basic purpose is to convert molecules that our bodies cannot use and cannot readily dispose of into more soluble, readily excreted derivatives. It is a physiological garbage disposal system. However, some molecules that have been modified by the P450 system become mutagens! One striking example is benzo[a]pyrene, an abundant constituent of tobacco smoke. Another example is aflatoxin, the product of a fungus that often infects grain and peanuts. The diagram shows how the complex ring structure of the benzo[a]pyrene derivative binds to DNA, creating a bulky adduct that can block DNA replication. The aflatoxin derivative is similarly bulky. The derivatives of benzo[a]pyrene and aflatoxin produced by the P450 system are well-known carcinogens (Lecture 6). The hydrazines and heterocyclic amines referred to in the previous paragraph also require activation by endogenous enzmes before they become carcinogens.

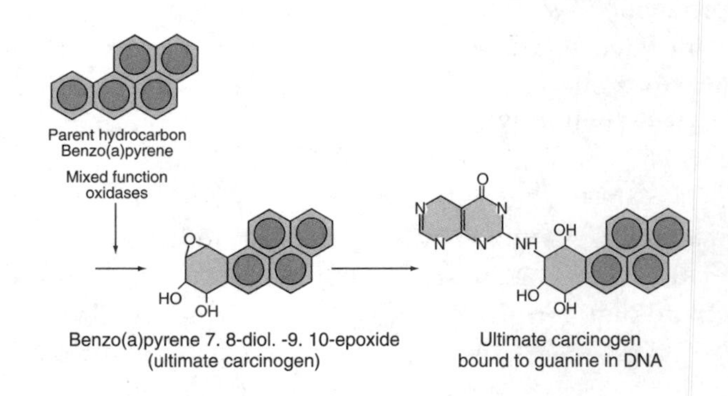

Figure 2-15 Modification of DNA by benzo[a]pyrene.

THE HUMAN MUTATION RATE

Although the word "mutation" properly refers to a change in DNA sequence, sometimes we use it to refer to a detectable variant phenotype. In either case, a basic concept is the locus-specific mutation rate, which is the frequency with which variants occur per generation in a given gene.

Nowadays most mutation detection is done by sequencing DNA. A common assay uses the *polymerase chain reaction (PCR)* to amplify all the exons from a gene. This very important technology is described in Box 2-1. The amplified exons are then sequenced and deviations from the normal are identified. Of course, this requires that the exon sequences be known, but as the Human Genome Project progresses, more and more genes meet that criterion. Not all mutations that cause disease are in exons, as was pointed out earlier in this lecture, so it is sometimes necessary to study a putative mutant gene in more detail. Actually, most of the mutations revealed by analysis of genes in persons with genetic diseases are not new mutations; alleles that were created several-to-many generations ago are responsible for most genetic disease. However, these assays also identify new mutations, particularly in the case of dominant genetic diseases (Lecture 3).

Prior to the availability of cheap and efficient DNA sequencing, attempts to measure the human mutation rate had to use less direct methods. The largest study on the human mutation rate was carried out for many years after World War II by the Department of Energy (DOE), to determine whether the children of persons proximally exposed to the atomic bombs in Hiroshima and Nagasaki carried more mutations than a control group. Their basic technique was *protein electrophoresis*, rather than screening for clinical abnormalities. Many changes in amino acid composition alter the size, the charge, or the configuration of a protein; such changes can often be detected by the rate at which a protein moves through a gel in response to an electric field. The DOE study, which was directed by the eminent geneticist James V. Neel, came to the conclusion that the locus-specific mutation rate was approximately 1×10^{-5} in both the exposed and control populations. People were relieved to learn that the descendants of the surviving victims of the atomic bombs in Japan would not have to pay a genetic price for their parents' misfortune.

Another way of expressing this result is that, for a given genetic locus, 1 gamete per 100,000 will have a mutation detectable by the methods used. If DNA sequencing had been available, it is obvious that more changes would have been detected. More recent studies have estimated the mutation rate per nucleotide pair to be about 1×10^{-8} per generation. This applies to noncoding, noncontrol sequence nucleotides, where changes usually will have no phenotypic effect. Coding and control region nucleotides presumably occur at the same rate initially, but many of them are not compatible with survival of the gamete or the embryo formed from that gamete, so they would not be detected in samples from living children.

Various other measurements of the mutation rates at specific genetic loci have given results that range from 10^{-4} to 10^{-6} per locus per generation. Numerous factors underlie this variation, among which are the size of the gene, its nucleotide content, the number of introns, and the frequency of repetitive elements within or near the gene (which will affect the probability of unequal crossing-over). We can combine the average mutation rate per base pair (1×10^{-8}) with the number of bp in the entire genome (3.2×10^{9}) to calculate that every gamete contains about 30 single-nucleotide mutations, and therefore every baby is born with approximately 60 such mutations. Most of those mutations, of course, will be phenotypically benign, but some may cause detectable abnormalities. Identifying the molecular basis of disease-producing mutations will be the subject of the next lecture.

BOX 2-1
THE POLYMERASE CHAIN REACTION

One of the most powerful techniques developed for molecular biology in the past two decades is the polymerase chain reaction (PCR). With PCR, short segments of DNA can be amplified to almost any desired extent. The target DNA can be a very rare component of a complex mixture of DNA fragments, such as a single exon from a whole genome. PCR can be thought of as a method for cloning DNA in a test tube, bypassing the complex process of making a recombinant bacterium.

The figure illustrates the basic steps in PCR. If the sequence of at least 20 nucleotides at each end of the DNA to be amplified is known, oligonucleotides that are complementary to those end sequences can easily be synthesized. A DNA sample containing at least one or a few copies of the specific DNA segment that is to be amplified will be denatured in the presence of the oligonucleotides and DNA polymerase. Then the mixture is cooled so that the oligonucleotides can anneal to the ends of the target DNA, and DNA polymerase then copies the target DNA, using the oligonucleotides as primers. This step doubles the amount of target DNA.

The process of denaturation, annealing, and DNA synthesis is repeated 20–30 times, with the target DNA being doubled each time. Twenty cycles will produce approximately 1 million times the starting amount of target DNA; 25 cycles will produce about 30 million copies of the initial target DNA. The nontarget DNA that was present initially is still there, of course, but it now is such a tiny fraction of the total DNA after PCR amplification that it can be ignored. One limitation of the PCR technique is that the DNA to be amplified cannot be very large; most PCR reactions amplify target DNAs that are only a few hundred bp long, but special contitions can increase the target size to several thousand bp.

Because each cycle in a PCR reaction requires denaturation of DNA with high heat, a special DNA polymerase that is heat resistant must be used. Typically, the enzyme used is prepared from a thermophilic archaebacterium named *Thermus aquaticus*, which lives in hot springs. The enzyme is called Taq polymerase.

The PCR reaction has been invaluable for the study of human genetic variation. At the clinical level, PCR makes it relatively easy to find the specific mutation in a patient who has symptoms of a known genetic disorder, by amplifying and then sequencing segments of the gene of interest. Knowing the precise mutation in a patient is often helpful in predicting the severity of the disease. In the future, it is quite likely that knowing the precise mutation will be helpful in designing specific molecular therapy for the patient. PCR also makes it easy to design sensitive tests for the presence of pathogens, such as the HIV or SARS viruses.

In the basic research context, PCR has been extremely helpful in understanding the molecular basis of many genetic diseases and in studying genetic variation in different human populations. The latter is helpful both in reconstructing the history of our species and in identifying the genetic factors that predispose to complex diseases.

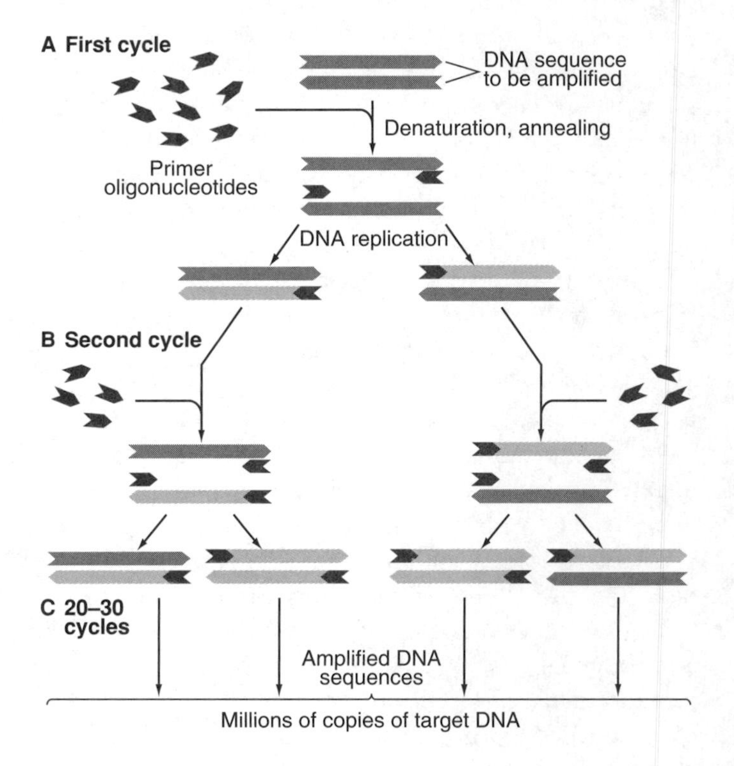

REVIEW

The major ways in which mutations change the human genome were summarized in this lecture. Most mutations are the consequences of normal metabolic processes—primarily errors in DNA replication, movement of transposable elements, and mistakes during meiosis, when chromosomes exchange material via the process of recombination. We call those sources of genomic alterations spontaneous mutations to distinguish their mode of origin from induced mutations, which are caused by DNA-modifying agents. The latter include free radicals, which are mostly produced as a byproduct of oxidative phosphorylation in mitochondria, and various mutagenic agents that enter the body via eating, drinking, or inhalation.

After a brief review of DNA replication, you learned that the major types of mistakes are incorrect incorporation (the wrong nucleotide is inserted) and replication slippage (random interruptions in base pairing between the template strand and the strand being synthesized may lead to inaccurate realignment of the two strands when replication resumes, especially in areas where there are short tandem repeats). One important class of genetic diseases arises from replication slippage in regions where there are long series of trinucleotide (triplet) repeats. We surveyed the effects of changes in base sequence on gene expression at both RNA and protein levels, introducing the concepts of missense, nonsense, and frameshift mutations.

We then mentioned that transposable elements occasionally cause genetic disease by integrating directly into a gene and affecting its expression. A much larger category of mutations arises from errors during meiotic recombination, when unequal crossovers may take place if two chromatids have misaligned, which usually involves repetitive elements. Chromosomal mutations were defined as microscopically visible changes in the number or structure of chromosomes; the major categories are aneuploidies, inversions, and translocations.

Induced mutations can be produced by ionizing radiation (such as x-rays or radioactive isotopes), ultraviolet radiation (sunlight), and chemical mutagens. The last category includes mutagens generated as normal byproducts of metabolism (principally free radicals) and mutagens that come from the substances we ingest or inhale. Some substances that are not toxic in their original form are converted to mutagens by the P450 enzymes, an unintended consequence of making them more soluble, so that they can be excreted. The lecture ended with some calculations of the human mutation rate.

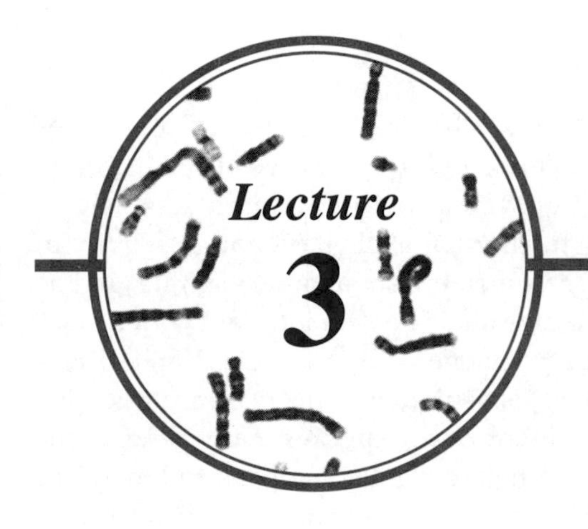

Genetic Diseases: The Consequences of Mutation

Mutations create genetic variation. Some variant alleles have no detectable consequences, some have good consequences, and some have harmful consequences. Good consequences include increased resistance to specific pathogens or parasites, as well as improvements in overall physical vigor and efficiency. Harmful consequences are the subject of this chapter, where we consider the role of genes in causing diseases, either directly or indirectly. We will begin with relatively simple situations, where it is clear that mutations at a single genetic locus can be responsible for a well-defined disease; then we will discuss complex conditions, where variations at one or more loci merely affect the probability that a clinically recognizable abnormality will develop.

MENDELIAN INHERITANCE OF AUTOSOMAL GENES

Gregor Mendel must have been an exceptionally clear thinker. While everyone else who was curious about inheritance was groping with half-baked theories, such as the idea that the characteristics of parents somehow blended like milk and water in the offspring, Mendel carried out his studies on pea plants with a *quantitative perspective.* When he had gathered data on the inheritance of seven easily observed characteristics (flower color, pea shape, plant form, etc.) he concluded that the results could be explained if the hereditary factors that determined each characteristic were discrete units, two in number in each individual, and that those units were passed intact from one generation to the next.

We must remember that Mendel didn't even know about the existence of chromosomes, let alone DNA or RNA and all the molecular relationships summarized in the first two lectures. His units of inheritance were theoretical concepts, and his rules describing the behavior of those units laid the foundation for *transmission genetics.* The Law of Allelic Segregation states that only one member of a pair of inheritable units enters a given gamete. We now realize that this law can be attributed to the segregation of chromosomes at meiosis. (If you didn't review meiosis while reading Lecture 2, you might want to do it now.) The Law of Independent Assortment states that alleles at different genetic loci assort randomly during the process of transmission to the next generation. Of course, we now understand that the process of distributing chromosomes at meiotic cell divisions doesn't distinguish chromosomes of paternal origin from those of maternal origin, so daughter cells get a random mixture of each type. Moreover, we also know that linked genes (genes close together on the same chromosome) do not assort independently. Mendel was either lucky, fortuitously choosing seven loci that were unlinked, or he had the good sense to ignore results that he couldn't interpret.

One more fundamental genetic concept came from Gregor Mendel. He observed that the hereditary units that controlled some of the plant characteristics he was studying would not appear in the offspring of first generation crosses, but would reappear in subsequent generations. Thus, a cross between white-flowering peas and purple-flowering peas would produce only purple-flowering plants, but when those first generation plants were allowed to interbreed randomly, about one-fourth of the second generation offspring would be white-flowering. From such observations, Mendel developed the concept of recessive and dominant inheritance, which applies to humans as well as to all bisexual organisms. Mendelian inheritance is an important factor in understanding human genetic disease and its consequences for reproductive decisions. Now, let's try to interpret Mendel's concepts in molecular terms.

In the simplest situation, a recessive allele is a genetic variant that has no effect on the phenotype of an individual if a normal allele is present. However, when both alleles at a given locus are recessive, a variant phenotype will occur, producing a recessive genetic disease if the function of that genetic locus is required for normal health. The twentieth century had barely begun when a remarkable Englishman named Archibald Garrod realized that Mendel's ideas could explain an unusual metabolic condition called *alkaptonuria* (urine that darkens rapidly upon exposure to air), which seemed to run in families, but did not occur in every generation. The existence of enzymes was known by then and Garrod postulated that alkaptonuria was caused by the absence of an enzyme. He was right; we now know the identity of the missing enzyme (homogentisic acid oxidase) and a variety of mutations in the gene have been characterized at the DNA level.

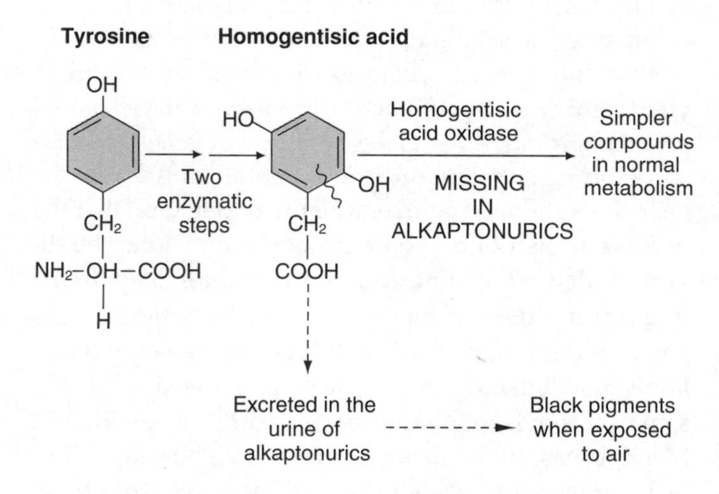

Figure 3-1 The metabolic block responsible for alkaptonura. From Mange, A. P. and Mange, E. J., 1990. *Genetics: Human Aspects*, 2nd ed, p394. By permission of Sinauer Associates.

Garrod extended his ideas to other biochemical abnormalities, and in 1909 published a seminal book, *Inborn Errors of Metabolism*, which eventually led to him being considered the father of biochemical genetics. Alkaptonuria is an excellent example of a recessive genetic condition, as are some of the better-known genetic diseases like cystic fibrosis, Tay-Sachs disease, and phenylketonuria (PKU), which have been the focus of intensive molecular analysis in the modern era. In all of these cases, the presence of one normal allele is sufficient for a normal phenotype, and the presence of the abnormal allele does not interfere with the function of the normal allele.

Recessive genetic diseases often involve genes that encode enzymes, and the recessive alleles are often null alleles (no enzyme activity produced). As long as the product of the normal allele has enough activity to perform its step in metabolism at a normal rate, and as long as the recessive allele doesn't interfere with metabolism, the presence of one recessive allele in an individual (a *heterozygote*) is benign. For this reason, a recessive allele may be present in a family for many generations before one of the heterozygotes happens to have children with another heterozygote and an *affected homozygote* offspring is born. The diagram below is a typical pedigree showing recessive inheritance.

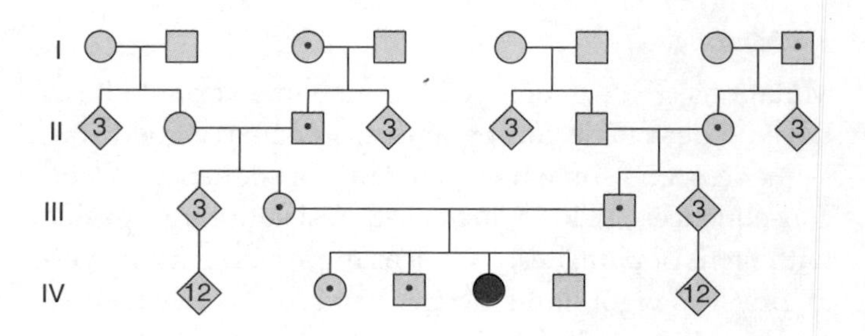

Figure 3-2 A hypothetical pedigree showing autosomal recessive inheritance of a genetic defect. Each generation is shown on a separate horizontal line. Circles ○ are females, squares □ are males, and diamonds ◇ are unspecified sex. Numbers inside symbols indicate the number of siblings not shown individually in the pedigree. Open symbols indicate homozygous normals, symbols with dots indicate heterozygotes, and filled symbols indicate homozygotes for the recessive allele.

Note that, in accordance with Mendel's first law, each heterozygote has a 50% chance of donating a recessive allele to any one child. Therefore, the probability that a particular child will receive both recessive alleles from his or her parents is $0.5 \times 0.5 = 0.25$. On a population basis, probability tells us that one-fourth of the children from matings between two heterozygotes will be affected; that is, they will be homozygotes for the recessive alleles. Of course, chance has no memory, so the odds of having an affected child are the same for each birth (one in four) and, given bad luck, some couples will have multiple affected children whereas other heterozygote couples blessed with good luck will have none. These facts are frequently misunderstood by the public. Couples who have had one affected child come in for genetic counseling, expecting to be assured that the next three children are guaranteed to be normal. Unfortunately, it doesn't work that way.

So at first glance, recessive inheritance appears to be simple. Yes, it *can* be simple, if you don't know much molecular biology. However, as we saw in the previous lectures, a typical gene contains hundreds of coding nucleotides and at least dozens of nucleotides in regulatory sites, and every one of them is mutable. Virtually every gene is potentially capable of having thousands of allelic variants, some of which will produce proteins that are functionally indistinguishable from the standard allele, some whose products have no activity at all, and others who will encode proteins with partially abnormal functions. Clinically recognizable genetic diseases are one end of a spectrum, where the most drastic effects on gene product function are found. If you look hard enough, you can almost always find people who are mildly affected, as well as people who only display abnormal function when subjected to unusual metabolic stress. Some alleles at a locus have no detectable abnormal effect unless certain other alleles are present elsewhere at other loci. We will get into that when we discuss multifactorial diseases.

But first, let's look at the other half of Mendelian inheritance in humans. Dominant genetic diseases occur when the presence of one variant allele is sufficient to create an abnormal phenotype. In pedigree analysis, dominant conditions are distinguished from recessive conditions if affected individuals usually have at least one affected parent and the offspring of matings between an affected heterozygote and a homozygous normal individual have a 50/50 chance of being affected and a 50/50 chance of being normal. The pedigree shown below illustrates autosomal dominant inheritance, such as one might see in a family where phenotypes such as brachydactyly (short fingers), Marfan syndrome or Huntington disease were present. The filled symbols indicate heterozygotes, but *in this case, heterozygotes are affected individuals.*

Figure 3-3 A pedigree showing autosomal dominant inheritance. From Vogel, F. and Motulsky, A. G., 1979. *Human Genetics: Problems and Approaches*, p.84. By permission of Springer-Verlag.

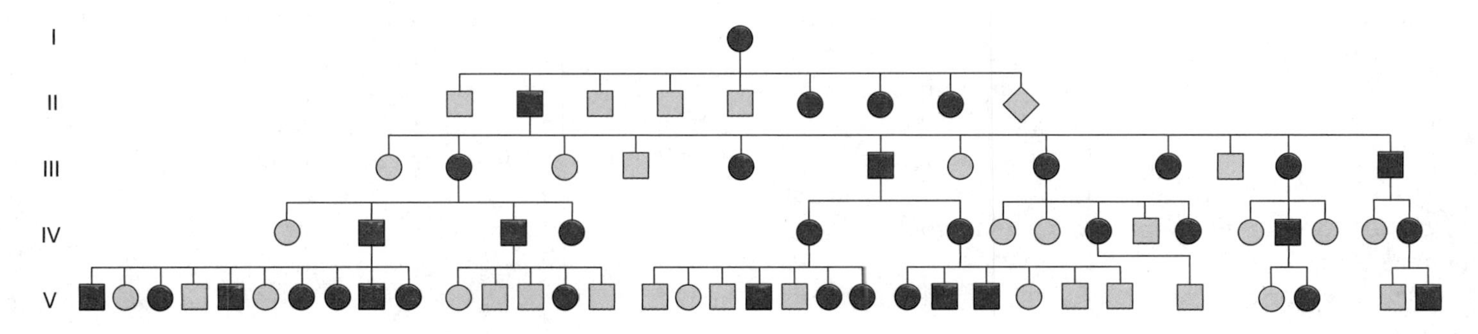

Dominant diseases have several molecular explanations; the major ones are haploid insufficiency, dominant negative effects, and gain-of-function mutations. Haploid insufficiency means that the presence of only one normal allele is not sufficient for normal function. In most metabolic pathways, there is a rate-limiting step; so, if a person has a null allele as one of the two gene copies at that locus, the activity of the normal allele may not be able to generate enough product to sustain a normal phenotype. A famous example occurs in the pathway for heme biosynthesis, where persons with one inactive allele in the gene for uroporphyrinogen synthase develop acute intermittent porphyria. The symptoms include colic, partial paralysis, and periods of mental confusion. It is believed that King George III of England suffered from acute intermittent porphyria. He was eventually declared incompetent and his oldest son served as regent for the remainder of the king's long life.

Haploid insufficiency may also involve genes that code for non-enzyme proteins, such as receptors and structural proteins. A well-known example is the gene that encodes the receptor for low density lipoproteins (LDL), which is crucial for the control of cholesterol levels in human blood. Mutations that reduce the number of functional receptors on the cell surface by 50% lead to the dominant disease, *familial hypercholesterolemia.* Persons with this condition develop atherosclerosis in early adulthood and are at greatly increased risk for heart attacks.

Dominant negative mutations lead to gene products that interfere with the function of the normal gene product. Most dominant negative mutations involve genes for multimeric proteins—proteins that have several to many subunits. If the presence of one or more abnormal subunits in a multimer reduces or abolishes the function of the entire multimer, then one mutant allele can be responsible for an abnormal phenotype. Marfan syndrome is a dominant genetic disorder that is caused by mutations in a gene for fibrillin, which is a protein that functions in long fibrous structures that are essential for the integrity of various organs and connective tissues. If half of the polypeptides that participate in forming fibrils are abnormal, the entire fibril may be abnormal. Individuals affected with the classic form of Marfan syndrome tend to be tall and thin, with various ocular and skeletal abnormalities; but the main threat to their lives comes from the tendency of the aorta to burst without warning. However, the fibrillin gene is known to have many variant alleles, some of which produce only mild symptoms. Again, the clinical disease is one extreme of a spectrum of phenotypes.

Gain-of-function mutations cause a gene or its product to do something that the normal allele does not do. Many gain-of-function mutations involve changes in transcriptional control, so that a gene is expressed in the wrong cell types and/or at the wrong stage of embryogenesis; the gene may or may not continue to be expressed in the usual cells at the usual times.

Another type of gain-of-function mutation underlies one of the most famous human genetic diseases—Huntington Disease (HD). It is a typical Mendelian autosomal dominant condition in its classic form. HD has received exceptional attention because of its symptomology. Abnormalities usually do not begin until the fourth or fifth decade of life, when involuntary twitching of the face or limbs appears. As time passes, these movements become more frequent and pronounced. Degeneration in the central nervous system results in irritability and finally dementia accompanied by physical helplessness. The time from diagnosis to death may be two decades. There is no treatment yet, so a diagnosis of HD is a death sentence of the most tragic type.

The HD gene encodes a large protein (huntingtin) whose normal function is not yet known. The gene contains a tandem series of CAG trinucleotides (glutamine codons) that are located near the amino terminus of the protein. Normally, there are less than 30 CAG repeats, but the disease-associated forms of huntingtin contain 36 or more CAG repeats. Although huntingtin is widely expressed throughout the body, the mutant protein appears to have a pathological effect in only a few cell types, especially neurons of the basal ganglia, which play a central role in the control of psychomotor behavior. Normally, huntingtin is a cytoplasmic protein, but the mutant form (or at least the poly-glutamine portion) accumulates in the nucleus of sensitive neurons. There it exerts its pathological effect, apparently through aggregation, although the details have yet to be discovered. Several other proteins have poly-amino-acid tracts that are responsible for neurodegenerative diseases when they are enlarged significantly beyond the normal size. This has led to the term *triplet-repeat diseases,* which was introduced in Lecture 2. Almost all of them are Mendelian dominants, and it is widely suspected that they are gain-of-function mutations.

HD is one of the few human dominant genetic diseases that can be described as a complete Mendelian dominant; homozygotes are not more severely affected than heterozygotes. Probably this fact is related to the gain-of-function property of HD mutant alleles; whatever the toxic effect may be, the expression of one such mutant gene is sufficient to do maximum damage. In contrast, most dominant genetic diseases belong to the haploinsufficiency or dominant negative classes, where the presence of two mutant alleles can usually produce a more severe phenotype than heterozygotes display. For example, homozygotes for Marfan syndrome or for achondroplasia (a form of dwarfism) die before birth, and homozygotes for familial hypercholesterolemia usually die from heart attacks before age 30.

Huntington disease also illustrates another interesting genetic phenomenon: It is the property known as *anticipation,* which means that the symptoms of the disease become more severe and have an earlier onset with each succeeding generation. Anticipation results from the tendency of the CAG repeats to increase in number each generation. The process is self-limiting, because children who are severely affected early in life will not reproduce.

Pedigree analysis of dominant disorders is often complicated by incomplete penetrance, the lack of symptoms in some individuals who are obligate carriers of the mutant gene, as exemplified by the next figure. Incomplete penetrance has multiple causes. For example, if Figure 3-4 represented a Huntington disease family, then individuals II-2, II-5, and III-10 might have died from heart attacks, accidents, or infectious diseases after their children were born, but before they lived long enough to display overt symptoms of HD. However, the usual explanation for incomplete penetrance is the presence of modifier genes, which are genes whose activity can change or prevent the development of abnormalities caused by a mutant gene. The figure could represent any autosomal dominant disorder for which there are skipped generations (i.e., presence of persons who have an affected parent and at least one affected child, but are apparently normal themselves).

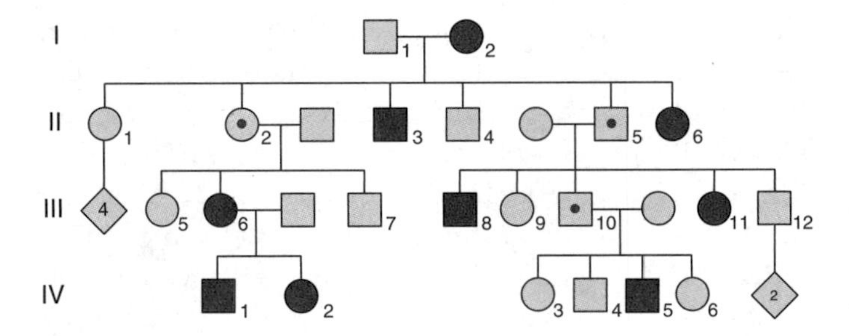

Figure 3-4 Incomplete penetrance in a hypothetical pedigree of a dominant disorder. Individuals indicated by circles with central dots must have carried the mutant allele, but did not express it. Adapted from Neel, J. V. and Schuil, W. J., 1954. *Human Heredity*, p50. By permission of University of Chicago Press.

A related effect is variable expressivity, the presence of different symptoms in individuals who must have the same mutant allele. Again, the best explanation is variation in the genetic background, but environmental factors can also lead to variable expressivity. We will visit these concepts later, in the section on complex inheritance. Also note that variable expressivity sometimes refers to phenotypic heterogeneity that is caused by different alleles at a given locus having various effects. For example, Becker muscular dystrophy and Duchenne muscular dystrophy are both caused by mutations in the giant dystrophin gene on the X chromosome; the milder, late-onset Becker form results from mutations that do not disrupt the reading frame of the mRNA, whereas the more severe Duchenne form of dystrophy is produced by nonsense, frameshift, and large deletion mutations.

MENDELIAN INHERITANCE OF X AND Y CHROMOSOME GENES

The basic facts about autosomal recessive and autosomal dominant inheritance also apply to genes on the sex chromosomes, but the quantitative relationships are different. Let's consider males first; they have one Y chromosome and one X chromosome in each somatic cell. The X chromosome is a medium-sized chromosome of 165 Mb containing about 1,500 genes, most of which have nothing to do with sex. The Y chromosome is smaller (60 Mb), with only about 50 functional genes, roughly half of which are involved with male sex determination or spermatogenesis. Near the tip of the short arms of both Y and X chromosomes there is a small area—*the pseudoautosomal region* (about 2.6 Mb)—where the same genetic loci are found on both chromosomes. Otherwise, men are *hemizygous* for genes on the X and Y chromosomes; that is, they have only one copy of each gene, instead of the usual two copies for autosomal loci.

Hemizygosity has interesting consequences. A mutation that behaves like a typical recessive in females will be phenotypically expressed in males. We call the phenotypes produced by such mutations *X-linked recessives,* even though they act like dominant conditions in males. One of the best-known X-linked recessive genetic diseases is hemophilia A, which is caused by mutations in the gene for Factor VIII, one of several proteins essential for blood clotting. In the past, men with hemophilia A rarely survived to reproductive age, because they would bleed to death from minor wounds during childhood. Nowadays, transfusions and/or replacement of Factor VIII, plus knowledge of their vulnerability, enables patients to have more-or-less normal life spans.

A very common X-linked trait is red-green color blindness, which is actually a cluster of phenotypes resulting from various rearrangements of two types of adjacent genes on the X chromosome (described in Lecture 2). Because these mild forms of color blindness have no detectable effect on reproduction, affected males can pass their mutant X chromosomes to their daughters, who are then usually heterozygotes whose vision will be normal. However, a mating between an affected male and a heterozygous female creates the possibility of homozygous female offspring, who will be color blind. The next figure illustrates these relationships.

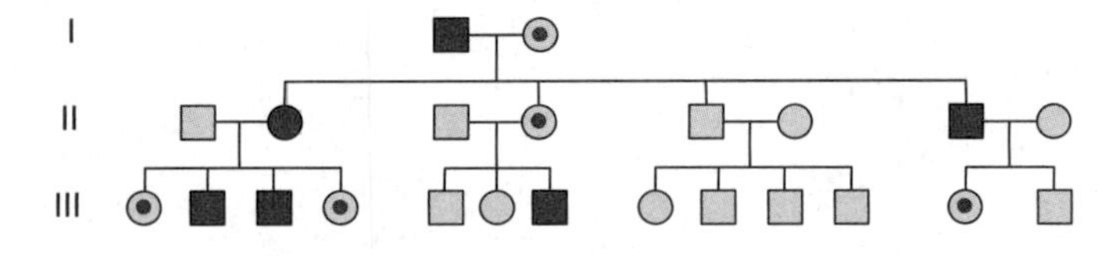

Figure 3-5 A pedigree of X-linked color blindness, showing how affected females can occur. Circles with central dots represent unaffected heterozygous females.

A fundamental functional problem arises from the fact that males have only one copy of the X chromosome and females have two. If the regulation of gene expression were the same on the X chromosome as it is on autosomes, females would make twice as much gene product as males make from every gene on the X. This would have devastating metabolic consequences, so some form of *dosage compensation* is necessary. Nature has solved this problem by inactivating one of the two X chromosomes in every somatic cell of females. Inactivation takes place at the blastocyst stage when there are 200–400 cells in a human embryo.

The choice of which X chromosome is to be inactivated is a random process, but once done, that X chromosome remains inactive throughout the rest of development and adult life, except during germ cell formation in females, when the inactivation is reversed. In interphase cells, the inactive X chromosome can be seen as a dark blob called a *Barr body,* against a granular background. Transcriptionally inactive regions of the genome often appear as condensed DNA-protein aggregates that stain darkly in cytological preparations, to which the name *heterochromatin* is given. Remarkably, in cells with an abnormally large number of X chromosomes, the number of Barr bodies is always one less than the number of X chromosomes. That's good from a functional perspective, because the normal situation is to have only one active X, but how does a cell count chromosomes? It's a fascinating puzzle.

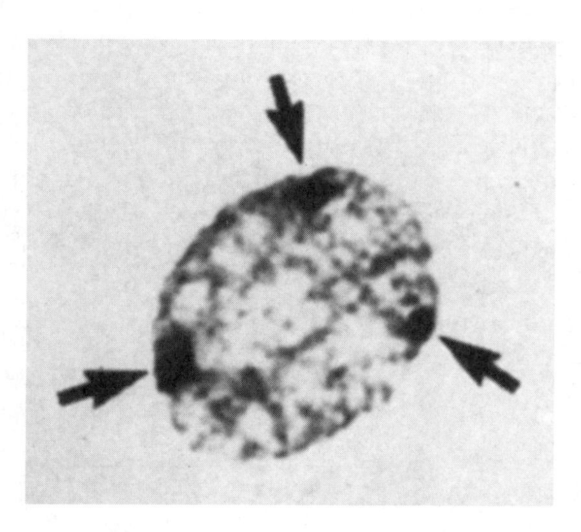

Figure 3–6 Three Barr bodies in a cell from a XXXX female.

The process of random inactivation of one X chromosome is called Lyonization, after Mary Lyon, who was the first to propose this mechanism of dosage compensation. Because Lyonization takes place in early embryos, rather large patches of adult tissues may have the same inactive X chromosome, because they are the descendants of a single blastocyst cell. If there is a mutant allele on one X chromosome in an individual woman, she will be a *functional mosaic* for expression of that gene. This can have effects on her health; examples include anhydrotic dysplasia, which is the patchy absence of sweat glands, and Rett syndrome, a neurological disorder that arises during early postnatal brain development.

Despite our detailed knowledge of the human genome sequence, we do not fully understand the molecular basis of X chromosome inactivation, although substantial progress has been made. It is clear that there is a locus called the *X inactivation center (XIC)* and that an RNA transcribed from the *Xist locus* within the *XIC* is crucial for inactivation. Xist RNA does not encode a protein; it is only synthesized from the inactive X and it only binds to that chromosome (that is, it is *cis-active*), which it eventually does at many sites as inactivation proceeds outward from the *XIC*. Transcriptional silencing cannot be initiated unless Xist RNA is available. Silencing apparently involves several proteins, some of which are modified histones, but the details are still obscure. A related gene is active only on the active X chromosome. It is named *Tsix*, because its RNA product contains the antisense sequence of the *Xist* gene. Recent research suggests that transcription of *Tsix* occurs before transcription of *Xist*, and that cessation of *Tsix* transcription on one X chromosome correlates with initiation of X inactivation via *Xist*. We don't know what gene controls expression of *Tsix*. All of this indicates that Xist and Tsix RNAs are very unusual molecules. Their properties are more than puzzling: They are mind-boggling.

Not all genes on the inactive X chromosome are inactivated. Many of them are found in the *pseudoautosomal region,* where recombination between X and Y chromosomes takes place during male meiosis or between the two X's in female meiosis, but a few other active genes are interspersed among inactive genes. However, the presence of that small set of active genes on both X chromosomes explains the fact that persons with abnormal numbers of X chromosomes are not fully normal phenotypically. Apparently, women with Turner syndrome (XO) have a deficit of one or more gene products, whereas Klinefelter males (XXY) and persons with more than two X chromosomes owe their problems to a surfeit of those gene products. The identity of the relevant genes is not yet known. There is evidence that a gene called SHOX in the pseudoautosomal region is at least partly responsible for the short stature that is one of the most visible characteristics of Turner syndrome, but it is likely that several genes are involved in producing the overall phenotype.

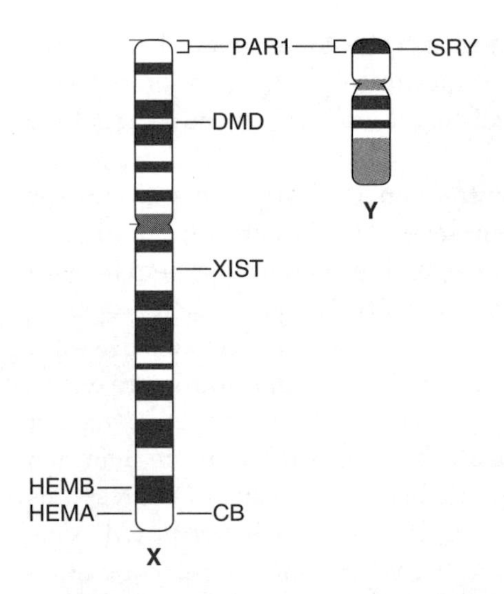

Figure 3-7 Comparison of X and Y chromosomes. PAR1—the major pseudoautosomal region; SRY—the testis-determining gene, just outside the PAR; DMD—Duchenne muscular dystrophy locus; XIST—X inactivation locus; HEMA and HEMB—hemophilia A and B loci; CB—locus for red-green color blindness.

The gene on the Y chromosome that has received the most attention is *SRY* (sex-determining region of the Y chromosome), which produces *TDF, the testis-determining factor.* It was identified a decade ago primarily by analysis of sex-reversed persons. Individuals who were genotypically XX but phenotypically male turned out to have small transpositions from the Y chromosome into one of their X chromosomes. Similarly, individuals who were genotypically XY but phenotypically female had lost a small portion of the Y chromosome. Putting all of those results together led to *SRY,* whose activity is required for establishment of maleness in humans. Originally expected to be a transcription factor, TDF (the *SRY* product) turns out to be a member of the HMG-box family, which binds to specific sequences of DNA and bends it at a sharp angle, bringing together nonadjacent sequences. Unidentified proteins then must interact with those DNA sequences to initiate a cascade of transcriptional events that leads to the male phenotype. A fascinating aspect of *SRY* is that its only nontestis site of expression is the brain. Cynics might conclude that this is molecular evidence supporting the folklore that a man's genitals control his brain, at least episodically!

FROM MENDEL TO GENES

Long before molecular biology revolutionized human genetics, scientists had identified the biochemical defect in several of the more common genetic diseases, such as *sickle cell anemia* and *phenylketonuria,* using educated guesses about what might account for the characteristics of those diseases. But more often, we really didn't have a clue about the basic biochemical defect or the gene whose abnormal function was responsible for the defect. Two of the best examples were cystic fibrosis (CF) and Huntington disease (HD): a classic Mendelian recessive and a classic Mendelian dominant disorder, respectively. HD was so mysterious that there weren't even good guesses available; CF had symptoms (buildup of mucous in the lungs, abnormally high salt in sweat, etc.) that suggested many hypotheses, but for a long time, all of them turned out to be wrong.

How do you find the gene that underlies a simple Mendelian disorder? Before we had the almost-complete sequence of the human genome there was a two-step strategy: *linkage analysis followed by positional cloning.* In other words, you first use linkage analysis to figure out the approximate location of that gene on a particular chromosome, then you look at clones representing DNA from that region until you find a strong correlation between the presence of a DNA sequence variant and the disease. The human genome sequence now supplements that basic procedure by offering candidate genes, but there are still many medically relevant genetic conditions for which the underlying genetic basis cannot be deduced from the phenotype, so linkage analysis is still very important.

Linkage analysis is based on the fact that genes located on the same chromosome are part of the same DNA molecule and they tend to be inherited together. In almost all organisms, each chromosome contains only one continuous DNA molecule, even though it may consist of hundreds of millions of base pairs. Genes on the same chromosome are said to be *syntenic* (on the same thread). Genetic linkage was not recognized until 1905, half a century after Mendel's pioneering work on the inheritance of seven characteristics in pea plants. Soon thereafter it was discovered that independent assortment was not true for all genes; alleles at two separate loci on the same chromosome were not always inherited independently. This led to the concept of linkage. Moreover, the combinations of alleles at different loci on one chromosome were sometimes rearranged during meiosis. Thomas Hunt Morgan in 1912 interpreted such phenomena in terms of exchanges between homologous chromosomes, for which he used the term, *crossing-over,* or in more modern terminology, *genetic recombination.* Morgan's name is now used for the unit of recombination.

Before we get into linkage analysis, let's review the basic facts about recombination (the topic was introduced in Lecture 2). During germ cell formation, specifically during prophase of the first meiotic cell division, homologous members of a pair of chromosomes undergo *synapsis* (i.e., they associate side-by-side in a very precise manner). At this time, each chromosome consists of two fully formed *chromatids*, joined only at their *centromeres* by special proteins; thus there are four chromatids in each pair of chromosomes (this was shown in Figure 2-8). Each chromatid represents a separate DNA molecule.

While the chromosomes are synapsed, crossing-over may occur between any two nonsister chromatids. If there are allelic differences at two or more locations in the homologous chromosomes, then the genetic result of crossing-over between two heterozygous loci will be *recombination* (see Figure 2-8). Recombination generates new combinations of alleles on each of the participating chromatids, but it does not ordinarily produce new alleles. Normally, crossing-over is perfectly reciprocal; there is neither loss nor gain of nucleotides in either chromatid. Unequal crossing-over sometimes occurs, particularly in areas where there are tandemly repeated sequences, as was mentioned in Lecture 2.

A single crossover involves only two of four chromatids in a synapsed pair of homologous chromosomes. Therefore, two recombinant chromatids and two nonrecombinant chromatids are produced by one crossover. If we consider the results of many meioses, either in one individual or in a population, the overall frequency of recombinant chromosomes is called the *recombination fraction* (RF).

If two genes on the same chromosome are close enough so that the recombination fraction is less than 50%, the two genes are genetically linked (i.e., alleles at the two loci do not assort independently). If two loci are syntenic (on the same chromosome), but so far apart that crossovers between them produce a recombination fraction of 50%, such genes are genetically unlinked, even though they are both physically part of the same DNA molecule that constitutes a given chromosome. In other words, when RF = 50%, we cannot distinguish genes that assort independently because they are on different chromosomes from genes that are on the same chromosome. The recombination fraction cannot exceed 50%, no matter how many crossovers occur between two loci. The explanation is too complex to present fully here, but the basic idea is that multiple crossovers often regenerate the original allelic combinations for the loci that are being assayed.

The overall frequency of crossovers in humans has been estimated in several ways. One important early study used the simple process of counting them in cytological preparations of spermatocytes; a crossover (a place where recombination is occurring) looks like an X or a fusion point when meiotic chromosomes are examined in the microscope. That study found an average of 52 crossovers per male meiosis (slightly more than two per chromosome pair). It is convenient to express this number in terms of *map units* or *centiMorgans (cM).* One genetic map unit or one cM is equal to a recombination fraction of 1%, over small distances. One crossover implies a genetic map length of 50 cM; that is, if two loci are separated by a distance such that an average of one crossover occurs between them in every meiotic cell, then those loci are 50 cM apart

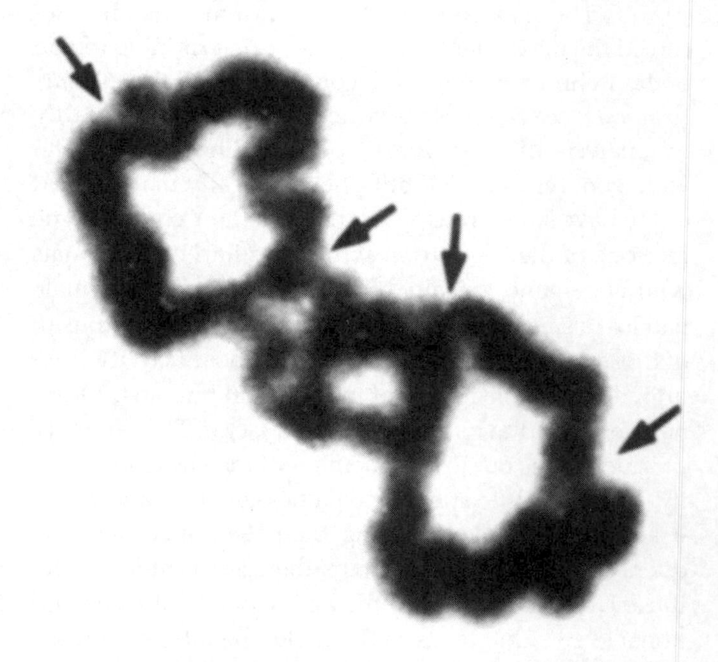

Figure 3-8 A pair of homologous human chromosomes, approaching metaphase of the first meiotic division. Arrows show four chiasmata (points where chromosomes appear to be joined), each of which implies that a crossover has occurred. (Courtesy of Maj Hulten.)

Extrapolating to the whole genome, we see that 52 crossovers implies a total genetic map length of 2,600 cM in humans. That number was expected to be an underestimate, because counting chiasmata on microscope slides is subject to several sources of error. More recently, the recombination frequencies between hundreds of pairs of marker loci have been calculated, and the current estimate is that the *genetic map for humans totals 3,500 cM*. The size of the genome is 3.2 billion bp, so on average, 1 cM is equivalent to about 1 million bp (1 Mb) in humans.

However, the quantitative relationship between frequency of recombination and the distance between genetic loci is not simple. There are regions where RF is much higher than the average and other regions where RF is much lower than the average. We don't know whether this variability is caused mostly by local details of base sequence, by the proteins that are bound to the DNA during meiosis, or by completely novel factors. A further complexity is that recombination rates are not the same in both sexes; they tend to be about one-third higher in females, although the sex difference is not constant from one region of the genome to another. This paradoxical phenomenon is not understood.

Linkage Analysis

The existence of genetic recombination is the basis for mapping genes by linkage analysis. If alleles at two loci assort independently, the loci are not close—they are either on different chromosomes or so far apart on the same chromosome that at least one crossover takes place between them in each meiosis, on average. However, if alleles at those two loci recombine rarely, they are tightly linked. By measuring the frequency of recombination between many pairs of loci, a genetic map can be constructed for each chromosome.

Genetic maps play an important role in the analysis of any genome. A genetic map gives the order in which genes occur in a chromosome, and the approximate location of a gene of interest. Genetic maps do not require knowledge of gene function; they can be constructed for genes that are known only from their effects on phenotypes. Genetic maps can also be constructed for variable DNA segments that are not part of genes, which has been essential in linkage analysis, as you will see presently. Although we have a nearly complete sequence of the human genome, that detailed physical map will not soon eliminate the usefulness of information on genetic linkage, for at least three reasons. First, we do not know the identity of many genes involved in rare genetic diseases; second, knowledge of the frequency of recombination in selected regions of the genome where important disease-causing loci occur will remain important for genetic screening and counseling (Lecture 4); and third, linkage analysis will often be crucial for determining whether a given clinical syndrome is caused by defects at more than one locus. The principles of linkage analysis are outlined next.

Let's suppose that we want to identify the gene responsible for *kramophilia*, an imaginary condition characterized by an uncontrollable fondness for junk food. We have done pedigree analysis and have conclusive evidence that kramophilia is a typical Mendelian dominant disorder. Where is the gene? To answer that question, we need a large series of *marker loci*, distributed throughout the genome. Then we can ask whether the *KPH* (kramophilia) gene is linked to one or more of those marker loci.

In the early days of human linkage analysis, there were few useful marker loci. A useful marker locus must be highly polymorphic, so that there is a reasonably good chance of a person with a disease gene of interest also being heterozygous at the marker locus. Having two different alleles at a marker locus is essential, because linkage analysis asks whether a particular allele at a marker locus tends to be inherited together with a particular phenotype within a family. The word *polymorphic*, which literally means "many forms," has received a special definition by geneticists: A locus is defined to be polymorphic if there are two or more alleles, each of which occurs with a frequency of 1% or more in a population. Thus, a locus with two alleles, one of which had a frequency of 99% and the other had a frequency of 1%, would qualify as a polymorphic locus by that definition. However, it would be of little use for linkage analysis, because only 2% of the population would be heterozygous. For linkage analysis, a locus must be polymorphic enough to generate at least 20% heterozygotes, and preferably more; otherwise, there would be little chance of finding families where both the disease and the marker alleles were segregating.

The first human linkage groups were identified by pedigree analysis in the 1950s, and all of them involved one or another of the *blood groups*, because they were the most polymorphic loci known at that time. The availability of highly polymorphic loci was dramatically increased when *restriction enzyme* digestion of DNA, followed by separation of fragments and identification of specific fragments, became a routine procedure. It soon became evident that some restriction enzyme cleavage sites were present or absent in different individuals. Next we learned that there are many series of tandemly repeated sequences in the genome (see Lecture 2) and the number of repeats at a given site is often highly polymorphic. The most useful type of tandemly repeated sequence turns out to be dinucleotide repeats, such as CACACACA. . . . , and we now have detailed genetic maps with the locations of hundreds of marker loci consisting of polymorphic short tandem repeats (STRs). We also have an increasingly dense map of single nucleotide polymorphisms (SNPs).

Now we are almost ready to map the kramophilia gene. We will choose a set of about 300 marker loci spaced about 10 cM apart. That way, the *KPH* gene must be no farther than 5 cM from one of those markers. The only other resource we need is a few *informative families*. What is an informative family? Take a look at the pedigree in Figure 3-9, which is a very informative family. For brevity, the pedigree uses the symbol K for the kramophilia allele and k for the fat-hating, fruit- and veggie-loving "normal" allele. The crucial individual is II-1; he is a heterozygote for K and k and also a heterozygote for a marker locus (any one of the 300 or so we have to test for linkage to *KPH*), with alleles M1 and M2. His mate, II-2, is also important. She is a homozygote at both the *KPH* locus (she has two k alleles) and the marker locus (she has two M2 alleles), which means that all of her gametes will be k/M2. If she were a heterozygote, she could make more than one class of gamete, and we couldn't figure out which alleles in the offspring came from which parent.

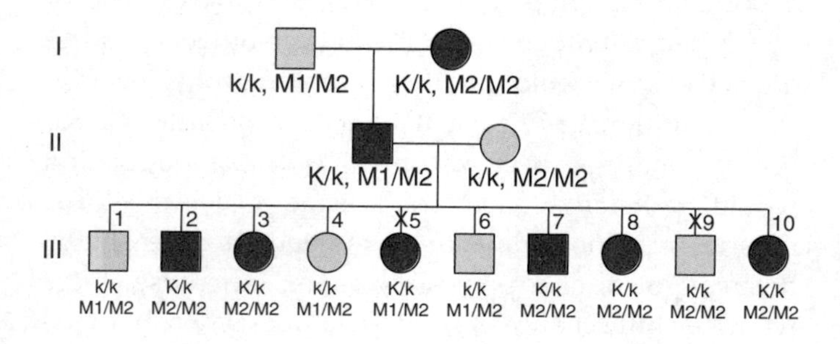

Figure 3-9 An informative family for linkage analysis.

There's more information regarding the family in Figure 3-9. Look at generation I, the parents of II-1. From their genotypes, we can deduce that II-1 received k and M1 from his father, whereas his mother contributed K and M2. That tells us the *linkage phase* of the alleles at both loci in II-2, which is basically just another way of describing which alleles came from which parent. Linkage phase is important, because if the two loci are linked, we can confidently identify recombinants in the third generation. That is, assuming that *KPH* and M are linked, we now know that the father of the 10 kids has k and M1 on one chromosome with K and M2 on the other member of that autosomal pair; so, if any of the offspring received K with M1 or k with M2 from dad, they must be recombinants. Finally, the family is big enough to give us some statistically useful results. In linkage analysis, the bigger the informative family, the better. Families with 10 or 12 kids are terrific, but several six-child families may be enough.

All right, is *KPH* linked to M or isn't it? What can we learn from Figure 3-9? We begin by asking two simple questions: (1) if the loci are unlinked, what is the probability of getting the ten genotypes in generation II; and (2) if the loci are linked at some specific recombination fraction (RF), what is the probability of getting the ten genotypes in generation III? Question 1 is easy to answer, because if the loci are unlinked, then the four types of gametes that II-1 can make (K/M1, K/M2, k/M1, k/M2) should occur with equal frequencies. That is, the probability of each gamete class is 0.25, and the probability of the entire set of children's genotypes for this family is $(0.25)^{10}$.

Question 2 is not so easy, because we have to calculate the probability of getting generation III at a specific RF. Look at the figure again and ask what were the genotypes of the gametes that came from II-1. You will find that eight of the children received either K/M2 or k/M1, the same combinations that II-1 got from his parents. One child (III-5) received K/M1, which implies a recombinant gamete; another child (III-9) received k/M2, which also implies a recombinant gamete. Two recombinant children out of 10 implies a recombination fraction of 0.2. Because there are two types of recombinant gametes, the probability of any recombinant gamete must be 0.1; similarly, the probability of any nonrecombinant gamete must be 0.8/2 = 0.4. Now we can calculate the probability of getting this family if RF = 0.2; it is

$$P(0.2) = (0.4)^8(0.1)^2$$

We are almost done. Now we want to ask, is the probability of this family occurring via nonlinkage greater than, equal to, or less than the probability of this family occurring if the K locus and the M locus are linked, with RF = 0.2? That's just a simple ratio, which we will call L, for likelihood:

$$L = (0.4)^8(0.1)^2 \;/\; (0.25)^{10} = 6.8$$

The formula states that the likelihood (L) or odds of getting the 10 children in the figure, given what we know about the genotypes of the parents, is 6.8 times higher for linkage with a recombination fraction of 0.2

than for independent assortment. What can we conclude? Well, geneticists have agreed that *the odds favoring linkage at a given distance to a marker relative to the odds of nonlinkage must be at least 1,000:1 before linkage is considered to be "proved."* Fortunately, data from many families can be combined (that is, the likelihood values can be multiplied, because each family is an independent event), so the odds we calculated in this example are not at all discouraging. Another convention is to convert the raw odds into a logarithm, giving a score called "lods," usually designated by the letter Z. Because adding logarithms is equivalent to multiplying ordinary numbers, *the lod score needed for proof of linkage is 3* (the common logarithm of 1,000). Our example has a lod score of 0.8.

Of course, linkage calculations are done by computers nowadays, and real situations are usually much more complicated than this example. However, this simplified account should have given you the basic concepts of linkage analysis, which is of fundamental importance for identifying the genes associated with human diseases.

Positional Analysis

To continue our pursuit of the mysterious kramophilia gene, let's suppose that linkage analysis has mapped the gene to a region of chromosome 19 covering 5 cM, or roughly 5 Mb of DNA, which is enough DNA to include at least 50 genes, if this happens to be a region with average gene density. We next consult the genome sequence database and ask what genes are in our region. We might find about 30 known genes and another 20 presumed genes (see Lecture 1). Next we look up the function of each known gene and ask ourselves whether one of them might affect metabolism by stimulating an abnormal craving for junk food; that is, are there any candidate genes in the region we are studying? If we find a candidate gene, we can focus all of our efforts on it until we either confirm it or reject it as the locus responsible for kramophilia.

Let's be unlucky and assume that there are no obvious candidate genes. Can we narrow the target region by using more closely spaced markers within the 5 cM area for linkage analysis? Maybe we can do that, but if the phenotype that interests us is rare, we probably won't have enough informative families to detect statistically significant differences in recombination rates with various nearby markers. One important strategy is to look for *linkage disequilibrium* between the abnormal allele at the *KPH* locus and one allele at a marker locus. Here's how that works.

Suppose there is only one mutation that causes kramophilia, and it originated several thousand years ago. At that moment, the K allele was associated with only one allele at each polymorphic locus on a particular chromosome in one person; it was in complete *linkage disequilibrium* with those loci. As the K allele was passed down from generation to generation, recombination would take place between K and distant loci on that chromosome, but for a very close marker locus, hundreds of generations might pass before a crossover occurred between K and the marker locus. Thus, at the present time K may be co-inherited entirely or almost entirely with only one allele at a nearby locus—the one that happened to be present on the chromosome in which the original K mutation occurred. We can look for linkage disequilibrium between the K allele and specific alleles at all available marker loci within the 5 Mb that linkage analysis identified as containing the *KPH* gene. The marker with the highest level of linkage disequilibrium would be a good starting point for more detailed molecular analysis.

What else can we do? There are several possibilities, but the general strategy is to find an association between a variation in gene structure or gene expression and the presence of the phenotype. The possibilities include absence of a given mRNA in affected persons, presence of an mRNA that is larger or smaller in affected persons than in unaffected persons, and presence of a DNA variant (a deletion, an insertion, or a base pair difference) only in affected persons. Lots of obstacles may have to be overcome. Some mRNAs, such as brain-specific mRNAs, may be available only from autopsies. Or, several to many nucleotide substitutions within the target gene may cause the abnormal phenotype, making correlation with a specific mutation impossible. It can be a long and difficult process, but eventually, the true kramophilia locus can be identified. Even then, we may be only part way to understanding the molecular basis of the condition, because the protein encoded by that gene may not have an obvious function.

ON BEYOND MENDEL: COMPLEX INHERITANCE

Many human traits display quantitative variation; examples include height, weight, skin color, and blood pressure. In most cases the variation is the net result of interactions between several to many genes and environmental factors (the environment includes everything that has happened to an individual since conception).

Sometimes more than one gene may affect the probability that a clinically recognizable condition will develop; sometimes environmental variables are the primary factors that determine whether a genetic potential for disease will be expressed; and sometimes both environmental and genetic factors are involved.

When a disease does not follow a Mendelian pattern of inheritance, the first question geneticists must ask is whether genetic factors play any role in causing the disease. We are all familiar with the expression that a particular condition tends to "run in the family," and epidemiological surveys that demonstrate the existence of *family clustering* are often the first good evidence that genetic factors may be involved. For example, 13% of children with one schizophrenic parent also have schizophrenia. This is 13 times the rate of occurrence of schizophrenia in the general population, which can also be expressed as a *relative risk* of 13. Relative risk calculations have provided evidence for genetic involvement in a wide variety of diseases and behavioral disorders, including several types of cancer, diabetes, heart disease, alcoholism, and bipolar disorder.

Adoption studies are another source of evidence for the role of genes in complex traits. Adopted children who have a biological parent with a given complex trait but whose adoptive parents do not have the trait are studied. If those children develop the trait more frequently than the general population, it is evidence for a genetic contribution to the causation of the trait.

Nevertheless, family clustering and adoption studies provide only indirect evidence for the existence of genetic factors. Fortunately, Nature has provided a simple tool for determining whether a condition is at least partially heritable: *identical twins*, which originate from a single zygote. The genomes of identical *(monozygotic or MZ)* twins are almost identical, differing only by post-conception mutations and DNA rearrangements (as in the immune system). The genomes of fraternal *(dizygotic or DZ)* twins are, on average, 50% identical, just like siblings in general.

If an abnormality occurs more frequently in both MZ twins than in both DZ twins, then it must be at least partly determined by genetic factors. If the *concordance rate* (both MZ twins are affected or not affected) is virtually 100%, the trait must be entirely or almost-entirely caused by genes. If the concordance rate for MZ twins is less than 100% but clearly higher than the concordance rate for DZ twins, both genetic and environmental factors are involved. If the concordance rate for MZ twins is not significantly higher than the rate for DZ twins, the trait is primarily a response to environmental variables.

How do we identify genes that contribute to the causation of complex disorders? The three main methods are *linkage analysis, allele-sharing, and allele-association.*

Linkage analysis can sometimes be applied to the analysis of complex traits if several large, multigenerational pedigrees are available. If a gene has a substantial effect on the probability of developing a given disease, it may be possible to locate that gene via linkage analysis, even though other genes also are involved in disease causation. One has to make guesses about dominant versus recessive inheritance and about the penetrance of the gene, but it can be done, given favorable family material. For example, linkage analysis was important in identification of the two breast cancer susceptibility genes, BRCA1 and BRCA2, even though 90% of all breast cancers are not caused by mutations in either gene.

The allele-sharing method is usually applied to sibling (sib) pairs, although other relationships can be used. One begins by identifying pairs of sibs (not twins) who are both affected with a trait; that is, they are *concordant.* Then a genome-wide scan for alleles at a large series of marker loci is conducted. If a marker locus is closely linked to the trait-related locus, the sibs are likely to share an allele at that locus. Look at the diagram below.

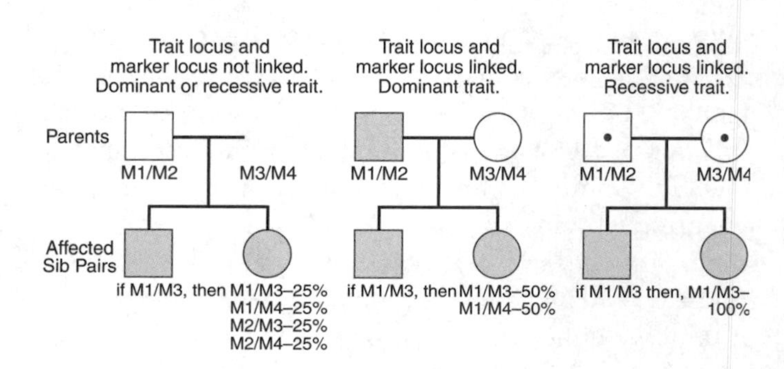

Figure 3-10 Sib-pair analysis.

Informative families at a specific marker locus will be those in which there are four distinct alleles in the parents. If the sibs received the same marker allele from both parents, that allele is said to be *identical by descent (IBD).* A sib pair may be IBD for zero, one, or two alleles at any locus. If the marker locus is not linked to the trait locus, then in a large sample of sib pairs, they will be IBD for zero, one, and two alleles in the ratio 1:2:1, as shown in the figure, left panel. If the marker locus is closely linked to the trait locus, then a series of affected sib pairs will have few or no cases that have zero IBD alleles, and the results will be different for dominant versus recessive inheritance of the trait. Various other factors complicate the analysis, such as recombination between the marker and trait loci. Of course, the basic fact that all complex traits may be caused by different combinations of variant genes at several loci implies that a very large sample of informative families may be needed before statistically significant conclusions can be obtained. Even so, the sib-pair method succeeds sometimes and is a useful procedure for the analysis of complex inheritance. Note that sib-pair analysis does not require linkage disequilibrium; the marker alleles that are IBD in one family may be distinct from marker alleles that are IBD in another family.

Allele-association studies are based on linkage disequilibrium, which was described earlier in this lecture. Once again, the basic idea is that when a wild type allele mutates to a form that increases the probability that a complex syndrome will develop, that new allele is in complete linkage disequilibrium with alleles at nearby polymorphic marker loci. For really close marker loci, linkage disequilibrium will persist for hundreds or thousands of generations.

Allele-association studies don't require family data. One begins by locating a large series (hundreds, at least) of *probands*—people who have the trait. They all donate a sample of their DNA (via blood or cheek swabs) and their DNAs are genotyped at all the marker loci; that is, a *whole genome scan* is performed. If one allele at a given marker locus tends to occur more frequently among the probands than among the general population, it indicates linkage disequilibrium, and a detailed search for genes in the vicinity of that marker locus can then be undertaken.

As with all the methods for identifying genes that contribute to the causation of complex traits, our ability to detect those genes is limited by the magnitude of the effect contributed by a given gene. The smaller the effect on causation, the larger the number of probands that must be examined. One common difficulty arises from the fact that more than one mutation at a given locus may cause the phenotypic effect; for example, if the effect is caused by reducing the activity of an enzyme or receptor molecule, there may be dozens of different mutations in the general population with that effect. You might wish to refer to Lecture 2, where we surveyed the many ways in which expression of a gene or the function of its product can become abnormal. Linkage disequilibrium cannot be detected in such situations. Therefore, allele-association studies are best conducted on populations that are more-or-less pure ethnically, so that the probability of there being only one disease-associated mutation is maximal.

Another problem, which is common to all the methods for identifying complex trait genes, is that diagnosis of the trait may not be uniform. This is especially true for behavioral disorders, where different investigators may have different criteria for classifying a person as affected or not (e.g., schizophrenia, bipolar illness). Those diagnoses depend on expert opinion, not on some biochemical test.

What have we learned above the causation of complex diseases? Let's look at the status of two intensely studied examples: diabetes and schizophrenia.

There are two types of diabetes. Type I is caused by total failure to produce insulin, which results from death of the beta cells in the pancreas. It occurs primarily in children and is widely believed to be caused primarily by a viral infection. Type II diabetes (also called maturity-onset or non–insulin-dependent diabetes) is characterized by resistance to the action of insulin, whose function is to control the conversion of glucose into glycogen within cells. As a result, blood sugar rises to a level that adversely affects many organs, causing circulatory insufficiency in the extremities, blindness, kidney failure, and a high risk of heart attack and stroke.

Type II diabetes is increasing at an alarming rate in the developed countries. In the United States, where at least 16 million people are overtly affected (and many others have early symptoms), there was a 49% increase in adults with diabetes between 1991 and 2000. This epidemic is not caused by an infectious agent; it is not the effect of sabotage by a foreign power or extra-terrestrials; it is the result of people eating too much, not getting enough exercise, and consequently becoming fat.

So, diabetes is fundamentally an environmental and behavioral problem. Do genes have anything to do with it? Yes, indeed, they do. We know that from twin studies, from the fact that not all obese people are diabetic and not all diabetics are fat, and from the striking differences in diabetes frequency between ethnic groups. In the United States, about 6.5% of whites, 10.2% of Hispanics, and 13% of African Americans have diabetes. The highest frequency appears to be in the Pima Indians of Arizona, where 50% of adults have diabetes, followed by the Oji-Cree Indians of Ontario, where 40% of adults are diabetic.

Extensive efforts to identify genes that contribute to the probability of developing diabetes are underway worldwide. At least 15 genes are on the list of suspects. Some of them are transcription factors that control insulin synthesis and secretion, some of them are involved with insulin signaling to target cells, and others affect the synthesis of hormones that interact with the function of insulin in various ways. Several of those genes have been identified as potential diabetes-causation factors by educated guesswork, looking at candidate genes because of what was already known about their functions. In other cases, linkage studies in extended families have identified a suspect gene. In the Pima Indians, a whole genome scan is being conducted. The overall picture that is emerging is an excellent example of the many ways in which disrupting a complex system can lead to the same clinical syndrome. Whether we shall ever understand the system well enough to offer every one lifelong freedom from diabetes, regardless of diet and lifestyle, is not yet predictable.

Schizophrenia (SZ) is one of the most common psychiatric disorders, affecting 1% of virtually all populations worldwide. It is characterized by a group of behavioral symptoms that fall into two classes. *Positive symptoms* include hallucinations, delusions, and bizarre thought patterns; *negative symptoms* include poor motivation, social withdrawal, affective flattening (subdued expression of emotions), and diminished executive function (difficulty making plans and carrying them out). Although the negative symptoms may make it impossible for a person with schizophrenia to hold a job or have normal social relationships, it is the positive symptoms that require treatment, because they may lead to violent behavior that becomes a danger to the patient or to others. These symptoms do not always occur together. There is a spectrum of abnormality ranging from mild personality disorders to full-blown SZ.

What causes schizophrenia? We don't know. The problem is more difficult than it is for diabetes, where elevated blood sugar directly led to the realization that either the synthesis of insulin or the body's response to insulin must be abnormal. Twin studies, family clustering, and adoption studies all demonstrate clearly that genetic factors are involved in susceptibility to schizophrenia. There is no shortage of metabolic differences between persons with SZ and the general population, but that is no surprise, given the complexity of the behavioral abnormalities. Numerous secondary and even less direct effects of the primary defect(s) are to be expected. In metaphorical terms, we don't yet know who is the criminal genius behind this problem (or set of problems). We are worse off than Sherlock Holmes, who was at least aware of Professor Moriarty.

The current clues to the genetic and metabolic causation of SZ have come from three sources. First, a group of drugs called *neuroleptics* has been very successful in controlling the positive symptoms of SZ, allowing many patients to resume more-or-less normal lives. *Typical neuroleptics* also produce side effects resembling Parkinson's disease, which is associated with degeneration of certain neurons that depend on dopamine, and this has led to the hypothesis that *dopamine* response is abnormal in SZ. *Atypical neuroleptics* (notably clozapine) also inhibit *serotonin receptors,* which suggests that serotonin response is abnormal in SZ. Other studies have implicated a third neurotransmitter, *glutamate,* or one of its receptors, in SZ.

A second source of clues to the origin of SZ comes from studies on whole brains, using MRI and other new techniques. These show an increase in ventricular size in SZ patients, especially in the prefrontal cortex and hippocampus, areas associated with cognitive functions and emotional regulation. There is also evidence that neurons in SZ brains are smaller, but not less numerous, than in control brains.

The third source of clues to the origin of SZ comes from genetics. Linkage studies have suggested at least eight loci that may confer increased susceptibility to SZ, but so far, no specific gene has been identified. Replication of linkage studies by different investigators has often failed, which may indicate either inadequacy of sample size or that SZ can be caused by interactions among several metabolic pathways. Allele-association studies are in progress, and one candidate gene, the serotonin receptor 5-HT2a, has been identified. One of its polymorphic forms appears to be associated with SZ, but the functional significance of that association has not yet been validated. Another candidate gene encodes catechol-O-methyltransferase (COMT), one of the enzymes that degrade dopamine and related compounds. The gene for COMT is located at 22q11, where a cytologically detectable deletion has been found in 2% of diagnosed schizophrenics.

GENOMIC IMPRINTING

Genomic imprinting is variation in the expression of a gene or group of genes that depends on whether those genes came from the male or female parent. The term *genomic imprinting* refers to the fact that the process of gametogenesis in one sex apparently marks some genetic material as being different from its counterpart supplied by the opposite sex. It is a major exception to the patterns of inheritance predicted by simple Mendelian dominance and recessiveness. At least 40 genes in humans are known to be imprinted, and many of them are associated with well-known clinical syndromes.

Early evidence for genomic imprinting came from work on mouse embryos in which both haploid chromosome complements were derived from the same sex parent. This can be achieved by nuclear transplantation. The male and female pronuclei remain separate for some time in the newly fertilized egg, and they can be distinguished from one another in the microscope. One pronucleus can be removed, and another pronucleus of the same parental origin as the one remaining in the egg can be obtained from another zygote and inserted into the first egg.

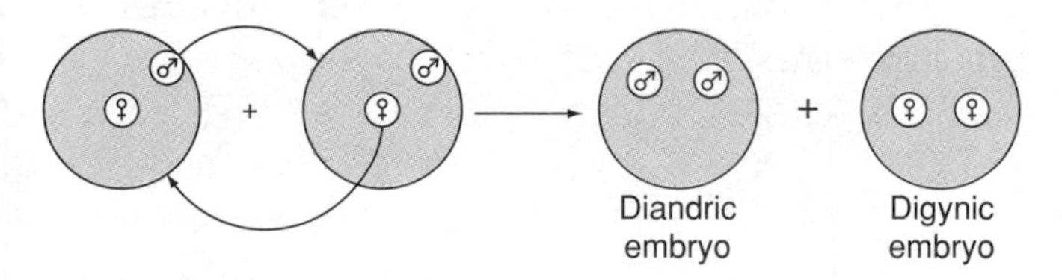

Figure 3-11 Production of diandric and digynic mouse embryos by transplantation of pronuclei.

Embryos with two maternally derived nuclei *(gynogenetic zygotes)* have poorly developed extra-embryonic membranes, although the embryos themselves are initially more-or-less normal. Embryos with two paternally derived nuclei *(androgenetic zygotes)* have relatively normal membranes initially, but abnormal embryonic structures. Both conditions are lethal. We are forced to conclude that both a maternal and a paternal pronucleus are necessary for normal development. An explanation for that conclusion is that some genes may be differentially inactivated *(imprinted)* during gametogenesis in the separate sexes.

In humans, Nature occasionally provides the equivalent of gynogenetic or androgenetic embryos for a single chromosome. *Uniparental disomies* are euploid genomes in which both copies of one chromosome or a major portion of a chromosome originated from one parent, the other parent having contributed nothing to that portion of the embryo's genome. In humans, uniparental disomies are believed to arise when a gamete that contains two copies of one chromosome (it is *disomic* as a result of nondisjunction during meiosis) joins with a gamete of the opposite sex that is *nullisomic* for the same chromosome (also due to non-disjunction).

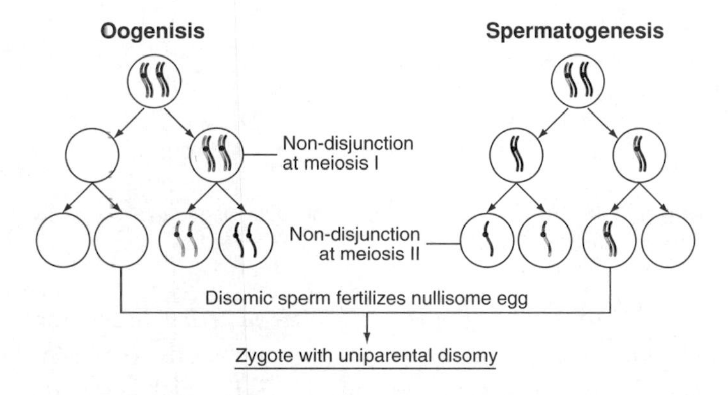

Figure 3-12 Origin of uniparental disomies.

Some well-studied examples of genomic imprinting are associated with *Prader-Willi syndrome (PWS) and Angelman syndrome (AS).* Figure 3-13 shows the pattern of inheritance for PWS, which occurs in about 1 in 10,000 births and is characterized by mental handicap and several minor physical abnormalities. PWS results from lack of expression of several genes in the area of chromosome 15q1, which are normally imprinted (i.e., inactivated) during female gametogenesis. Thus, only the paternal alleles are active in normal persons. If a deletion occurs in 15q1, there will be no genes expressed from that region in children who receive the deletion from their father, because the maternal alleles have been silenced by imprinting. Conversely, all of the children of an affected mother will be normal (assuming a normal mate), but some of her sons may have affected children of either sex.

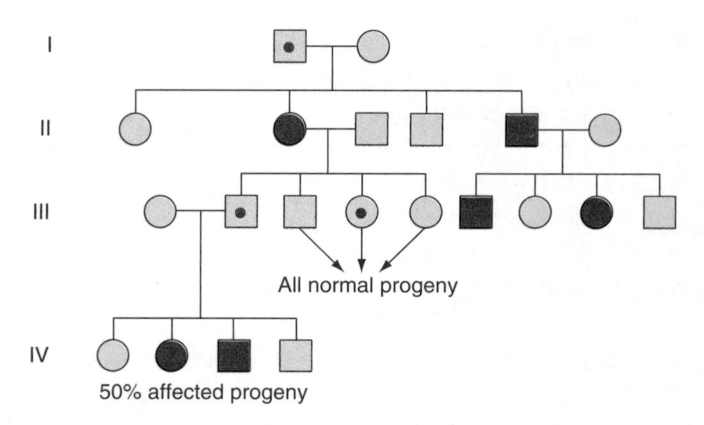

Figure 3-13 A pedigree showing incomplete penetrance resulting from genomic imprinting.

In some patients with PWS, *maternal disomy* for chromosome 15 has been documented. Conversely, AS can also be caused by a deletion of 15q1, which is microscopically indistinguishable from the deletion that causes PWS. Nondeletion AS can be the result of *paternal disomy* for that portion of chromosome 15. The two syndromes are clinically distinct. It is likely, therefore, that there is differential imprinting of a cluster of genes in the 15q1 region, and that both a paternally derived chromosome and a maternally derived chromosome are required for normal development.

The fundamental mechanism involved in imprinting certain genes appears to be methylation of DNA, especially in transcriptional control regions. In vertebrates, methylation of DNA is almost exclusively found on cytosine residues, where a methyl group can be attached to the carbon atom at position 5 by a methyltransferase enzyme (see Figure 2-2). The methyl groups occur mostly on cytosines that are followed by guanines; these are called CpG dinucleotides. Regulatory sequences of genes that are actively expressed are usually unmethylated or hypomethylated, whereas in tissues in which a gene is not expressed, the regulatory sequences are often heavily methylated.

Imprinting is a cyclical process. At some time during the life of primordial germ cells, a wave of demethylation occurs, wiping out the patterns that have been in place throughout the life of that individual. New methyl groups are added to imprintable genes during gametogenesis (or no later than the pronuclear stage in zygotes). The signals for this *de novo* methylation must be a property of the DNA sequence at specific sites, perhaps with the collaboration of DNA-binding proteins other than the methylases. Once a methylation pattern has been created, it is usually maintained in all the descendants of a zygote, via the action of enzymes called *maintenance methylases.* Their function is to place methyl groups on newly replicated DNA, using the methylation pattern of the parental DNA strand as a template.

In most cases, imprinting that is established in the gamete is permanent and affects all somatic cells of the adult. There are a few exceptions where imprinting is tissue specific. An example is Angelman syndrome, where the gene UBE3A is imprinted only in certain portions of the central nervous system. Only the maternal allele is expressed; the paternal allele is imprinted. Therefore, if an inactivating mutation is present in the maternal allele, there will be no gene product made from that locus in the imprinted portions of the brain. This leads to the neurobehavioral abnormalities that characterize AS. In the rest of the body, the paternal allele will be expressed and half the normal amount of the *UBE3A* gene product appears to be sufficient. In general, haploinsufficiency could complicate the effects of mutations at a gene that is imprinted in a tissue-specific manner, leading to a variety of abnormalities in locations other than the imprinted tissue. Indeed, some very complex and variable phenotypes are associated with other imprintable genes.

REVIEW

This lecture began with a review of Mendelian inheritance, which describes the fundamental rules of transmission genetics. The basic aspects of autosomal recessive and autosomal dominant inheritance were summarized, from the point of view of human genetic diseases. Recessive conditions frequently result from inheritance of two null alleles—genetic variants with no enzyme activity or no protein product. Dominant conditions result from haploinsufficiency (half the normal amount of a gene product is not enough), dominant negative mutations (the product of a mutant allele interferes with the function of the normal allele's product), and gain-of-function mutations (a gene product does something new and deleterious). Many dominant genetic diseases display incomplete penetrance, which is absence of symptoms in persons who must have a mutant allele, as shown by pedigree analysis. Incomplete penetrance is usually an effect of other genes that vary from individual to individual and impinge on the function of the mutant gene.

Human males have one X and one Y chromosome, rather than the two X chromosomes present in females. This accounts for the fact that only one copy of a mutant allele of a gene on the X chromosome is enough to produce an abnormal phenotype in males, even though the same allele may be recessive in females. Early in embryogenesis, one X chromosome is largely inactivated in each cell of a female, but the choice is random. A few X chromosome genes on the inactive X escape inactivation; many of them are in the pseudoautosomal region, most of which is located at the tip of the short arms of both X and Y chromosomes. The presence of two active copies of genes in that region is necessary for normal function. The Y chromosome contains only a few dozen active genes, some of which are required for the establishment of maleness and the production of sperm.

I then described the basic strategy for identifying a gene responsible for a disease that is inherited according to the rules set forth by Mendel. An approximate chromosomal location for the gene is established by linkage analysis, which is a process of correlating the co-inheritance of the disease and a nearby genetic marker locus, such as a short tandem repeat, within large families. The suspect region is then subjected to fine-scale positional analysis by several methods, one of which is to search the genome sequence for genes of known function that might account for the disease phenotype. If that fails, an attempt to find linkage disequilibrium between the disease phenotype and a nearby marker locus is made; this is a search for specific marker alleles that are so close to the unknown disease-associated locus that recombination between them almost never occurs. Thus, only one allele at the marker locus is co-inherited with the disease. Linkage disequilibrium data will often identify the disease-associated gene.

The next section of the lecture dealt with the problem of identifying genes that contribute to disease causation in a quantitative way, but do not individually determine normality versus abnormality. Behavioral disorders, diabetes, and cardiovascular disease are well-known examples of complex diseases whose pattern of inheritance does not fit the standard Mendelian rules. First, the major techniques for finding out whether genes play a significant role in causing a complex disease were summarized; these include family clustering, adoption studies, and twin studies. When it has been established that a condition is at least partly related to heritable factors, the genes that contribute to heritability are sought by several methods, such as sib-pair analysis and allele-association studies, which require large samples of affected persons and sophisticated statistical analysis. Diabetes and schizophrenia were described as examples.

Genomic imprinting was our last example of non-Mendelian inheritance. This is a process by which a few individual genes are marked (usually during gametogenesis) in such a way that only one copy will be expressed during most of embryogenesis and adult life. For some genes, the maternal copy is inactivated; for other genes, the paternal copy is inactivated. Mutations in genes that are normally imprinted may lead to genetic disease, with unusual inheritance patterns. The detailed mechanism of imprinting is not fully known, but it involves methylation of specific DNA regions. Imprinting is wiped out during germ cell formation. The biological function of imprinting is not yet understood.

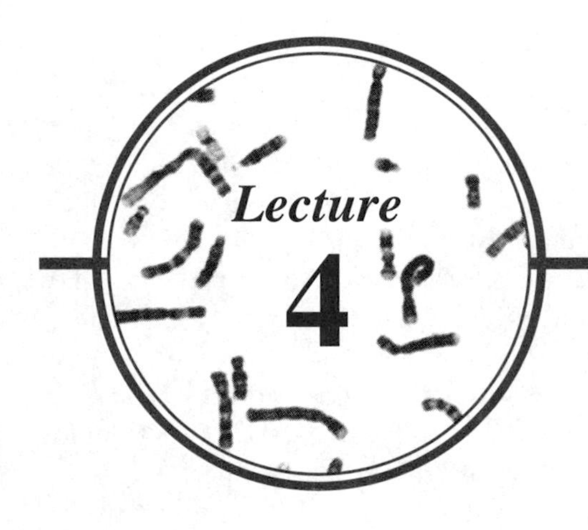

The Human Genome and Medical Practice

Diseases and other abnormal conditions that are primarily caused by single genes are present in about 1 in 100 births in the United States. Complex diseases (such as diabetes, heart disease, and cancer), in which genetic variants act as predisposing factors, are responsible for a major portion of morbidity and the majority of deaths. Genes also affect our sensitivity to various infectious and parasitic agents, both through the immune system and through variations in many other molecules, such as cell surface receptors. All together, a major fraction of the patients seen by clinicians express symptoms that are either directly or indirectly related to their genomes. Recent progress in analyzing the human genome has made it much easier to recognize the contribution of genetic variations to disease causation and in some cases to design optimal treatments. We are in the very early stages of individualized medicine and the best is yet to come.

Before a disease can be treated or cured, it is necessary to recognize its presence and its identity. For adult patients, this is usually the result of a sick person seeking medical advice. For newborns, whose metabolic abnormalities may not be apparent at birth, genetic screening programs have been extremely beneficial, so we will begin this lecture with a survey of genetic screening. Treatment of disease, whether the disease is caused by genetic factors entirely or in part, obviously involves the entire panoply of clinical methodology, from antibiotics to surgery. We will look at some successful nutritional therapies for genetic disease, then we will continue with a section on stem cells, which offer exciting potential for organ repair and replacement. In addition, there are now intensive efforts to find ways of replacing and/or repairing defective genes, which is the topic of gene therapy. The lecture will end with a brief description of the emerging field of pharmacogenomics and the prospects for designing optimal treatment for individuals based on their genotypes.

GENETIC SCREENING

Target populations for genetic screening vary widely. Newborn genetic screening currently focuses on detection of single-gene disorders. Adult genetic screening is often done at the family level, in order to identify couples at risk for birth of an affected child. Screening is sometimes done on specific adult groups in which there is a relatively common mutation that significantly affects the probability of developing a complex disease, such as breast cancer. In the future, it may become possible to screen every newborn for large numbers of polymorphisms that influence a person's susceptibility to a variety of complex disorders, including cardiovascular disease, arthritis, and diabetes.

Newborn Genetic Screening

The goal of newborn genetic screening is to identify abnormalities before they become symptomatic, while they are susceptible to preventive treatment or at least to treatment that will lessen the severity of symptoms that would otherwise develop. Screening programs are only carried out for disorders that are treatable. Genetic screening of newborns began in 1962 in Massachusetts, using a simple test developed by Robert Guthrie, and has now spread to all parts of the United States as well as many other countries. Guthrie developed his test in order to identify infants with abnormally high levels of *phenylalanine*, which often is indicative of the absence of *phenylalanine hydroxylase (PAH)*, an enzyme that converts phenylalanine to tyrosine. Infants lacking PAH develop *phenylketonuria (PKU)* within a few weeks, leading to severe mental retardation and a set of other neurological abnormalities. The Guthrie test uses a mutant bacterial strain to assay for phenylalanine in a drop of blood obtained from an infant's heel about 24 hours after birth. If an abnormally high concentration of phenylalanine is present, the bacteria grow conspicuously, as illustrated in Figure 4-1.

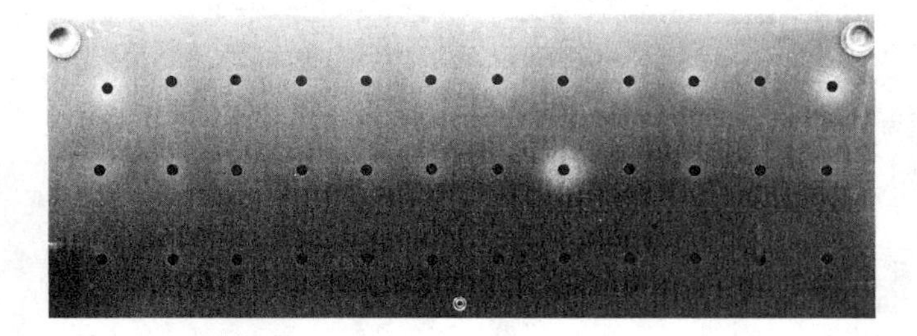

Figure 4-1 Part of a Guthrie test plate, showing 36 samples. Notice the bacterial growth halos in the corners of the top row, which were controls, and the fifth spot from the right end of the second row. The latter came from a child subsequently diagnosed as having PKU.

Children with classic PKU (blood phenylalanine about 10 times the normal level) require lifelong institutionalization, which is an enormous cost to society and/or their families. The incidence of PKU in the United States is about 1 in 10,000 live births. The Guthrie test is cheap and can be performed on thousands of blood samples in one laboratory. Not only is it fully justifiable on economic grounds, it avoids the personal suffering of patients and their families. Children who are identified as potential PKU patients are placed on a low phenylalanine diet. If they follow the diet throughout childhood, they will have average intelligence when adult.

Not long after screening for PKU came into general use, it was realized that the same approach could be applied to several other treatable and relatively common genetic disorders that also make diagnostic changes in the amount of easily assayable metabolites in blood. Screening programs for *congenital hypothyroidism (CH)*, *galactosemia*, *homocystinuria*, and *maple syrup disease* are now widespread. Assays for sickle cell anemia, biotinidase deficiency, and congenital adrenal hyperplasia can also be done on the blood samples from neonates. Among these, the most successful program has been the detection of congenital hypothyroidism, which is almost completely curable by oral administration of *synthetic thyroid hormone*.

Biological variability creates several technical problems for Guthrie-type newborn screening programs. Let's first consider something that we discussed in Lecture 3, which is that *defects in more than one gene may lead to the same clinical phenotype*. An example is that a small fraction of infants with hyper-phenylalaninemia have nothing wrong with their phenylalanine hydroxylase; they have a genetic defect that leads to lack of a cofactor (*tetrahydrobiopterin or* BH_4) for PAH. Treatment for those infants is quite different from treatment for classic PAH deficiency. So, when it has been established that a particular infant has an abnormally high level of phenylalanine, it is prudent to do a secondary assay for BH_4 deficiency.

An even more general problem is false positives and false negatives. Because of inherent biological variability, not all infants with genetic defects in phenylalanine metabolism will have blood levels of that amino acid that are clearly distinct from the normal level at the time the blood sample is taken. In order to avoid missing them, it is necessary to set the sensitivity of the assay at a level that leads to some *false positives*—apparently high phenylalanine levels that are not actually indicative of PKU. The false positives can be identified by doing a second assay on infants with borderline levels of phenylalanine. This is not always easy logistically because it requires that the parents bring the child back to the hospital or other clinic where another sample can be taken. Moreover, it is frightening for the parents. False positive rates for newborn screening for various genetic diseases range from less than 1 per 1,000 to roughly 1 per 100.

The opposite problem is *false negatives*—failing to identify a child who really does have a genetic defect. The most common source of false negatives is laboratory errors, but biological variability is also involved, because some infants do not accumulate the diagnostic metabolite until after they have been released from the hospital (usually one or two days after birth). The best way to deal with the latter situation is parent education, so that infants displaying any symptoms that seem abnormal will quickly be brought to the attention of a physician.

Screening for multiple metabolites in Guthrie-type blood specimens from newborns is becoming more sensitive and effective with increasing use of a sophisticated new technique known as *tandem mass spectrometry*, which is capable of assaying for at least 25 genetic disorders with a single assay. As the name suggests, the molecular masses of various metabolites in a dried blood sample are measured by a mass spectrometer. The word "tandem" refers to the fact that the mass measurements are first done on intact metabolites, and then on fragments of the metabolites produced by collision with an inert gas under high pressure. The technique is quite sensitive, which alleviates some of the false negative problem; it is also quite precise, measuring more than one product of a metabolic pathway, which alleviates some of the false positive problem. The full potential of tandem mass spectrometry for newborn genetic screening has yet to be determined, but the technique is clearly a significant advance over one-at-a-time assays.

Adult Genetic Screening

There are no nationwide adult genetic screening programs at present, but several possibilities have been considered. The best candidate is *hemochromatosis*, an autosomal recessive disorder of iron metabolism that usually does not become clinically evident until midlife, although an elevated iron level in blood may indicate a predisposition to the disease much earlier. Homozygotes for susceptibility to hemochromatosis are quite common: 1 in 200 to 500 Caucasians. For unknown reasons, men become symptomatic five times more frequently than women. If untreated, the disease causes life-threatening damage to the liver, heart, pancreas, and other organs.

Effective treatment for hemochromatosis is simple: avoid iron supplements (many multi-vitamin tablets contain iron) during the early stages and undergo *phlebotomy* if iron overload develops. One cannot ignore the irony (pun intended) of this situation. Long ago, doctors withdrew blood from their patients to treat almost every disease, usually with deleterious consequences; now in this era of molecular medicine, we have a disease where bleeding the patient is the best treatment!

The fact that effective treatment for hemochromatosis is available is essential background information before any population-based screening program could be considered. How would persons genetically predisposed to hemochromatosis be identified? The disease is caused by mutations in a protein called *HFE*, which binds to a cell surface protein, the *transferrin receptor*, and participates in the regulation of iron uptake into cells. Two mutations in the HFE gene account for at least 70%, and probably more, of all mutant alleles, so a DNA-based assay would be easy to design. It has been estimated that up to 1,000 lives could be saved each year in the United States if everyone were screened for those mutant alleles. Nevertheless, government agencies have recommended against it, citing a variety of reasons including incomplete penetrance (not all persons who are susceptible to the condition will become ill), stigmatization, and discrimination. *Stigmatization*—the perception by individuals and/or their peers that having a mutant gene makes them inferior—is a serious problem among less educated segments of the population, but it will diminish in importance as genetic medicine becomes more familiar. *Discrimination*, especially by insurers, is discussed later in this lecture.

Population-wide screening for mutant alleles that predispose to hemochromatosis, if it were to be established, could be combined with screening for many other mutant alleles, such as those that affect a person's probability of developing breast cancer, colon cancer, heart disease, and diabetes, to mention only the more common disorders. All of these tests would be intended to inform the public so that they could reduce their chances of becoming ill, usually by modifications of lifestyle, but developing an adequately informed public might take generations. And who would have access to the test results? Would every physician who sees a person in the clinic, for whatever purpose, have access? Many physicians would feel that access was essential, if they were to offer every patient comprehensive preventative care.

Another category of adult genetic screening is family-based screening. In contrast to population-based screening, where the emphasis is on the health of the persons being screened, the usual reason for family-based screening is to avoid the birth of a child affected with a serious genetic disease. *Cystic fibrosis* (CF) is a well-known example. This autosomal recessive disease is caused by defects in a chloride channel protein known as the *CF transmembrane regulator (CFTR)*. It affects chloride and sodium ion transport across the apical membranes of epithelial cells, primarily in the airways of the lungs, sweat glands, and pancreas. The result is that mucous normally produced by these cells is thickened, leading to clogging of the airways and pancreatic ducts especially. The thick mucous in the lungs also encourages bacterial infections, and CF patients need a lot of antibiotics. Progressive deterioration of lung function usually causes death by age 30, although the severity of symptoms varies widely.

Cystic fibrosis occurs in about one in 1,500–2,000 births in populations of Western European descent, which means that approximately one person in 20 in that ethnic group is a heterozygous carrier of a mutant allele. Couples in which one or both partners have a relative with CF often want to know whether they are at risk for having an affected child. If they are, they may elect to have no children, to adopt children, or to have their own children. Couples that already have a child affected with CF are vitally interested in learning whether the next child will be normal. That information can usually be obtained by *pre-implantation genetic diagnosis (PGD)* on embryos conceived by in *vitro fertilization (IVF)*. PGD uses one of the first few cells from an embryo for a DNA-based assay (via PCR) for a particular mutant allele. Because IVF routinely produces several viable embryos, the physician can screen all of them and choose a normal embryo for implantation into the mother.

Of course, not every couple who knows they are at risk for a child with a serious genetic disease wants to undergo IVF and PGD. It is expensive and emotionally stressful. Some may elect to wait until a pregnancy is established, then have amniotic cells from the embryo assayed for a mutant allele that is known to occur in one of the parents. (Note that it is not necessary to test for both parental mutant alleles when dealing with a recessive disease; one normal allele is enough for a normal phenotype.) Most couples who choose to have an at-risk fetus assayed intend to abort the fetus if it is affected, but some just want to be informed prior to birth, so that they can be prepared to deal with the child's special needs immediately.

It is worthwhile to pause here and ask whether a population-based screening program for cystic fibrosis would be beneficial, assuming that problems of stigmatization and discrimination could be kept within tolerable limits. The arguments in favor of such a program are that the disease is common, and that caring for an affected child is enormously expensive in material terms and immeasurably costly in emotional terms. Unfortunately, there's a practical problem that is caused by a fundamental aspect of genetic diseases: there's more than one way to inactivate a gene. The famous deltaF508 mutation (mentioned in Lecture 2) in the CFTR gene is responsible for 70% of the mutant alleles in some populations of Western European descent, but it is much less frequent in ethnic groups from southern Europe, and quite rare elsewhere. Overall, more than 800 mutations in the CFTR gene have been shown to be associated with CF worldwide! If population-wide screening for CF alleles were to be done, it would not be feasible to include all of them in the assay. Then, what do you tell couples who have a child affected with CF after having been told that they do not have any of the mutant alleles that are responsible for 95% of CF cases, "Tough luck, folks—you can't say we didn't warn you"? No matter how carefully the genetic test results were worded, it is certain that there would be misunderstandings in abundance and some lawsuits. The same problem applies to population-based screening for hemochromatosis and nearly all other genetic diseases.

Even so, the American College of Medical Geneticists, the American College of Obstetricians, and the National Institutes of Health recommended in 2001 that CF mutations with a carrier frequency of at least 0.1% in the overall United States population should be screened for, especially in adult relatives of patients with CF and in partners of patients with CF. It turned out that 25 mutations met that criterion; a test that screens for all of them is now available commercially, and some doctors recommend it for any couple planning a pregnancy. So far, the biggest problem that has arisen has been misunderstanding of the results.

All of this section's examples dealt with disorders that are inherited according to Mendelian rules and occur in a very small fraction of the population. Increased knowledge of the genes that contribute to complex disease causation is likely to lead to new genetic screening programs in the near future. You may want to review the section of Lecture 3 where we discussed complex diseases, because this is really an extension of that topic. Here's the basic idea: We know a few situations where specific genetic variations increase the probability of acquiring a disease, but do not act in a straightforward Mendelian manner. One well-known example is allele 4 of the gene that encodes apolipoprotein E (APOE). Numerous studies have shown a higher frequency of this allele in Alzheimer's patients than in controls, and it is generally agreed that APOE-4 increases the risk of developing Alzheimer's disease, although the extent of the increased risk is controversial. APOE plays a role in cholesterol transport, but how this could contribute to Alzheimer's disease is still unknown.

As we learn more about genetic contributions to the predisposition to develop diseases such as atherosclerosis, rheumatoid arthritis, diabetes, and other common afflictions (Lecture 3), we will develop a catalog of disease-related alleles at many loci. There is a tremendous amount of variation in the human genome, and most of it takes the form of single nucleotide polymorphisms (SNPs, pronounced "snips"), where a single base pair in some individuals differs from the base pair found in most people at that location. In fact, there is approximately one SNP per 1,000 bp (1 per kb), which means your genome differs from your best friend's genome at roughly 3 million places. Most SNPs are in noncoding regions, and although a few of those presumably affect gene regulation, the vast majority must be functionally inert. However, a small percentage of SNPs occur in coding regions (in exons), some of which are *synonymous changes* (they don't change the encoded amino acid, because of the redundancy of the genetic code) and some create changes in the amino acid specified at that site. The nonsynonymous coding SNPs are a likely source of variations in protein function that contribute to the probability of developing various complex diseases.

In 1999, 10 large pharmaceutical companies and the Wellcome Trust in Britain, recognizing the potential importance of SNPs for the analysis of human gene function in both health and disease, as well as for drug development, created the *SNP Consortium*, with the goal of developing a vast database of human SNPs that would be available to the public. They succeeded beyond expectations, and several million SNPs have now been identified, which is approximately 1 SNP for every 1 kb in the human genome, on average.

As we develop a catalog of disease-associated SNPs, it will become feasible to assay every person's genome for hundreds (or thousands) of them, thus making available a tremendous amount of information that individuals can use for lifestyle decisions and their physicians can use for individualized treatment decisions.

Population-based SNP screening programs will employ the new technique of DNA microarray analysis, which already offers the possibility of assaying for thousands of genes or mRNAs at once, so SNP surveys that seemed impractical a few years ago will soon be feasible at an acceptable cost. This is a fundamental advance in the tools available to genetic researchers as well as clinicians, so pause here and learn about the technique from Box 4-1 (page 67). Clinicians are currently using this technique to characterize different kinds of tumors in terms of gene expression patterns or to define a person's response to medication in great detail; there are also many applications to basic research.

Microarrays—also known as DNA "chips," because they are manufactured with techniques borrowed from the semiconductor industry—can certainly be used for SNP assays, and several biotech companies are actively working on instruments and techniques to make high-throughput SNP assays maximally efficient. Think about it this way: Each spot on a SNP chip will have an oligonucleotide, a piece of single-stranded DNA about 25 bases long (a 25-mer), with the central base being one that is polymorphic in the population. We can probe that chip with small pieces of fluorescent DNA from the genome of any person, healthy or not. We can adjust the hybridization conditions so that, if the test DNA contains a sequence precisely complementary to the 25-mer on the chip, a fluorescent DNA-DNA hybrid will be formed; but, if there is one base different from the 25 bases in that oligonucleotide on the chip, no hybrid will be formed. The microarray can contain thousands of 25-mers, each of which contains a sequence that represents a different SNP site.

Because one SNP chip can theoretically provide an encyclopedic amount of information about a person's genome, it is likely that there will be many proposals for population-based genetic screening programs. Inevitably, there will be advocates for screening all newborns, and others who prefer to limit screening to adults. Some will urge mandatory screening (especially for newborns), and others will prefer strictly voluntary programs. It will take a long time to sort out the competing interests.

Undoubtedly, there will be powerful commercial pressures to establish mass screening programs, because of the huge profit potential. It is unfortunate that those pressures may influence decisions that should be made only after extensive and objective consideration of the individual and social benefits and liabilities that may ensue when any new screening program is sanctioned by the government or by the medical profession. In addition to the problems already mentioned, we cannot escape the fact that mass screening for alleles that underlie genetic diseases or predisposition to complex diseases has eugenic overtones. Especially if a mandatory screening program were established, the mere existence of the program would suggest to many people that society had decided that it is wrong to be born with any of those diseases or susceptibilities. The negative aspects of genetic screening programs cannot be completely eliminated, but it is to be hoped that positive aspects will eventually dominate, both at the individual and policy-making level. After all, the goal of screening is to offer better health to as many people as possible.

Ethical and Legal Aspects of Genetic Screening

Acquisition of information about the genotypes of individuals and populations has serious and complicated ethical and legal ramifications. People tend to perceive genetic information as much more personal than standard clinical data, such as blood pressure or cholesterol level. Thus, the existence of genetic data inevitably creates the possibility of an invasion or reduction of privacy. Ethicists point out that *privacy has both intrinsic and instrumental value*. Intrinsic value implies that privacy is an essential component of our humanity. Without privacy, there cannot be autonomy, the freedom to make choices about the direction and quality of one's own life. The instrumental value of privacy refers to the possibility of genetic information being used either for harm or for good, either to individuals or to populations.

Probably the most actively discussed aspect of genetic privacy relates to insurer access to genetic information. In countries with universal health insurance, the problem is not serious. However, it is an acute problem in the United States, with its chaotic system of private health insurance for most persons of working age (usually funded through employers), partial insurance via Medicare for the elderly, and no insurance at all for at least 15% of the population. Americans correctly fear that their genetic information may have an adverse effect on their insurability. Insurers acknowledge this, pointing out that they cannot remain in business if they offer insurance to individuals whose medical costs are likely to be greatly in excess of the premiums paid.

Discrimination by insurers based on genetic information is less irrational now than it was several decades ago, when misunderstanding of the medical significance of genetic test results was commonplace. However, the implications of genetic information are only rarely definite, especially with regard to complex disorders. A person with a particular allele may be more likely to develop heart disease, diabetes, or cancer than a person with a different allele, but so much depends on the rest of each person's genotype and on lifestyle choices, that predictions of morbidity are highly uncertain. As our knowledge of the interactions among genes and environmental variables increases, we will be able to make better predictions of outcomes, but a substantial level of uncertainty will always be present.

Most states have enacted genetic nondiscrimination laws, but they vary widely and often have major loopholes. Usually the laws prohibit genetic testing or disclosure of genetic information to a third party without informed consent, although exceptions are made for anonymous research or law enforcement. Paradoxically, genetic information is usually defined as coming from direct tests on DNA, whereas information relating to the protein products of genes is unregulated. Another way in which insurers and employers may gain access to genetic information is via benefits records (reimbursement for specific medical conditions), which are not classified as confidential, in contrast to the results of medical examinations.

The possibility of genetic discrimination by insurers poses a dilemma for individuals, who may forego testing that could help them deal with serious health problems in order to avoid the possibility of losing their insurance or of becoming uninsurable. Whether it will be possible to design a system that provides virtually fool-proof security for genetic information is far from clear, but it is essential to try.

THERAPY OF GENETIC DISORDERS

Some form of therapy is available for virtually all genetic disorders. In many cases, therapy is essentially palliative, such as physiotherapy to loosen lung secretions and antibiotics to control infections in cystic fibrosis patients. Another example would be analgesics for pain management in sickle cell disease patients, often combined with blood transfusions. Examples for the application of virtually any aspect of clinical medicine to the amelioration of symptoms caused by genetic diseases could be found, but the focus of this book is on the genes and gene expression, so we will limit our survey to treatments that are directly aimed at genes and at the primary biochemical defects resulting from mutations in specific genes.

Nutritional Therapies

There are a few genetic diseases for which essentially normal health can be achieved by altering the patient's diet. We have already mentioned the example of PKU, where restricting the intake of phenylalanine from infancy through childhood (or throughout life) avoids most of the impairment of intellectual capacity that would otherwise occur. Congenital hypothyroidism has no adverse effects when synthetic thyroid hormone is taken as a dietary supplement daily. Removal of galactose from the diet prevents the symptoms of galactosemia. Control of iron intake, sometimes supplemented with phlebotomy, controls hemochromatosis. Careful control of fat intake improves the prognosis of patients with familial hypercholesterolemia.

Many readers will be familiar with the heartwarming story of *Lorenzo's Oil,* which is a nutritional treatment for *X-linked adrenoleukodystrophy (X-ALD).* This is a lipid-storage disorder; metabolic products known as very long chain fatty acids accumulate in cells of the brain and adrenal gland especially, leading eventually to failure of those organs and death. The severity varies greatly as does the age of onset, but even boys who will be severely affected do not show symptoms before four years of age. As many as one male child in 20,000 may be affected.

When the parents of Lorenzo Odone were told that their 5-year-old son had X-ALD and would not survive more than two years, they undertook an intensive study of the relevant biochemistry and neurology, eventually concluding that a dietary supplement consisting mainly of olive oil and rapeseed oil might be an effective treatment. Their son showed some improvement, and in 1992 a movie named *Lorenzo's Oil* was released, telling their story. Unfortunately, Lorenzo's condition has slowly deteriorated, but he still survives. In the meantime, investigators have learned *the oil is most effective if given to susceptible boys before symptoms appear*. Recent results indicated that as many as two-thirds of those boys may remain normal if they take Lorenzo's Oil faithfully. It is not yet known whether the protective effect of the oil will last throughout a normal life span, but there are now apparently normal young adult males with mutations in the *X-ALD* gene.

Protein Replacement Therapies

Many genetic diseases create morbidity because a protein needed for normal health is absent or nonfunctional. Almost any protein can now be manufactured in recombinant organisms, so supply is no longer the primary obstacle to designing protein replacement therapies. The primary problem is: How do you get the protein to the place in the human body where it is needed, in sufficient quantities to cure the metabolic defect? Most proteins are intracellular. If a recombinant protein is injected into the blood, it is likely to be degraded by proteases in a brief time, and repeated injections may generate an immune response that can be fatal.

If a protein survives its journey through the blood and is taken up by a target cell, it usually enters *lysosomes,* where a complex array of degradative enzymes soon put an end to it. Of course, the enzymes that live in lysosomes are designed to be stable in that environment, and that bit of natural history has led to the successful development of replacement therapy for several *lysosomal storage diseases*. This is a group of genetic disorders that arise from the absence of one of the degradative enzymes that normally function in lysosomes. The molecules they would ordinarily degrade accumulate within lysosomes, leading to cell enlargement and eventually to cell death. The severity of the effect varies from one cell type to another, and with the exact nature of the missing enzyme.

Gaucher disease is the prototypical lysosomal storage disease, caused by the absence of *glucocerebrosidase*. It is an autosomal recessive disorder; heterozygotes for one mutant allele are phenotypically normal. Although the enzyme is widely expressed in the body, *macrophages* are responsible for nearly all of the disease phenotype (there is a very rare form of the disease that affects neurons). Unhealthy and dead macrophages cause enlargement of the liver and spleen as well as bone marrow lesions, along with anemia and platelet deficiency. Symptoms in adults vary from mild discomfort to severe illness, sometimes leading to death in the third decade, but the age of onset varies from early childhood to old age. This variability is believed to be due to the particular alleles present in an individual as well as to the genetic background (the rest of that individual's genotype).

Glucocerebrosidase initially was purified from pooled human placentas, modified to make it more likely to be taken up by macrophages, and injected into the blood of Gaucher patients, where it was taken up into macrophage lysosomes—its normal environment. More recently, an enzyme produced by recombinant DNA techniques has become available. Both treatments reduce the pain and other symptoms of Gaucher patients significantly. However, the enzyme must be injected at least every two weeks, and the annual cost of treatment exceeds $50,000 per year. Many patients who would benefit from enzyme replacement don't get treated. Whether adequate amounts of recombinant enzyme can ever be produced for a small fraction of the current cost is not yet known.

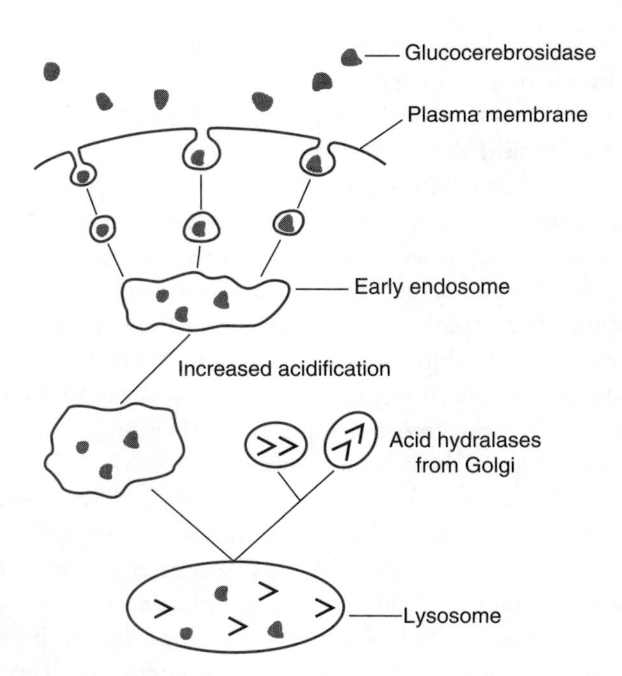

Figure 4-2 Endocytosis of lysosomal enzymes.

Gaucher disease is particularly common in *Ashkenazi Jews* (Jews from Eastern Europe) and their descendants, where the frequency of heterozygotes may be as high as one in 15. Only a handful of alleles account for 95% of the mutations in that population, so genetic testing is being offered routinely in some areas, along with tests for *Tay-Sachs disease* (another lysosomal storage disorder) and several other genetic disorders that have relatively high frequency among the Ashkenazim. For a variety of reasons, geneticists suspect that heterozygotes for recessive alleles at the Gaucher and Tay-Sachs loci may have had a selective advantage in the ghettos of Eastern Europe, but what that advantage might be has not yet been deduced.

Protein replacement therapy can also be effective when the missing or defective protein is a normal constituent of blood. The *hemophilias* provide an instructive example. The inheritable nature of the tendency to bleed excessively in response to wounds and trauma that would be quickly healed by blood clotting in most people was recognized many centuries ago when the authors of the *Talmud*, a rabbinical commentary on Jewish laws, recommended that circumcision of baby boys should no longer be required in families where two male children had already bled to death.

Blood clotting is a complex process that involves more than a dozen proteins, most of which are proteases or cofactors of proteases. Two of the genes that encode clotting factors are located on the X chromosome; they are responsible for the vast majority of hemophilias. Hemophilia A affects 1 male child in 5,000; it results from defects in factor VIII, an exceptionally large protein. Hemophilia B affects 1 male child in 30,000; it results from mutations in factor IX. Both proteins have essential roles in *conversion of prothrombin to thrombin*, a protease that converts *soluble fibrinogen to insoluble fibrin*, the main component of blood clots. Because factors VIII and IX affect the same process, the phenotypes produced by defects in either factor are inseparable at a gross level. Of course, with modern molecular methods, it is not difficult to find out exactly what molecular lesion is present when a child with hemophilia develops symptoms.

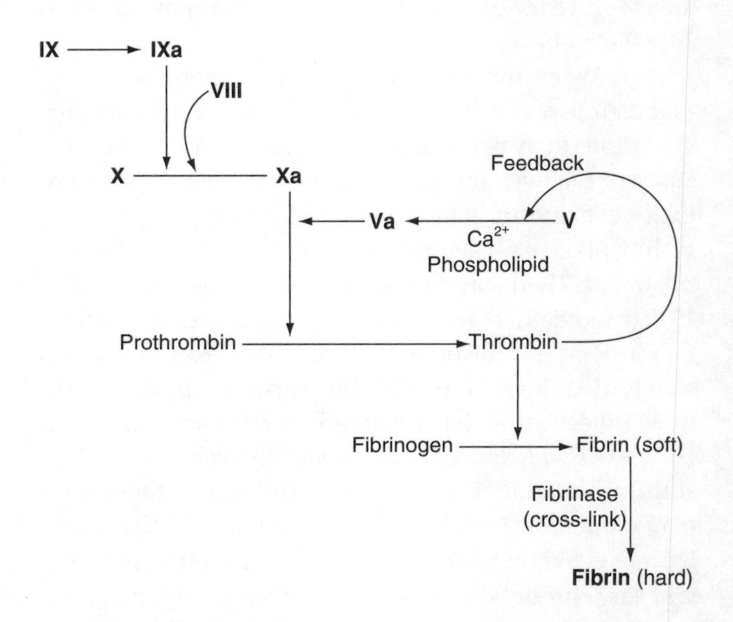

Figure 4-3 Some relevant steps in blood clotting.

The most famous sufferer from hemophilia was the Tsarevich Alexis, heir to the throne of Russia, who was assassinated at the age of 14, along with the rest of his family, during the Bolshevik revolution. Alexis was a descendant of Queen Victoria of England, who apparently was a carrier of a new mutation. Through her daughters, the mutant gene was disseminated to several royal houses. It is not known whether Victoria was a carrier of hemophilia A or B, and because there may not be any surviving carriers of the mutation among her descendants, the diagnosis may never be made.

Treatment of hemophilias began decades ago with blood transfusions, which were supplanted by plasma concentrates of coagulation factors in the 1970s. At first, it was routine for those concentrates to be contaminated with hepatitis B or C viruses. The situation became far more serious in the 1980s when about two-thirds of hemophilia patients in the United States became infected with HIV. Much safer plasma concentrates are now available, as are factors VIII and IX prepared from cells genetically engineered with recombinant DNA.

However, protein replacement therapy for hemophilias is not an uncomplicated success, because *many patients develop neutralizing antibodies to the injected proteins*. This should come as no surprise, because we all know that one function of the immune system is to recognize foreign proteins and synthesize antibodies to those proteins as part of the process of eliminating them. In patients whose mutations lead to the production of little or no normal factor VIII or IX, their immune systems react to exogenous clotting factors as foreign proteins. For example, if a nonsense codon occurs near the 5′ end of a gene, full-length polypeptides won't be made. On the other hand, if the mutation is a missense codon, the protein will be made, even though it contains an amino acid substitution that inactivates it, and the immune system will regard that protein as a normal constituent.

The *alloantibody problem* can be addressed by genetic engineering methods, because some of the most antigenic portions of a clotting factor are not necessary for its function. Those portions can be deleted from the recombinant factors used for treatment of hemophilia patients. Those approaches may not solve the problem for all patients, so there is great hope that gene therapy will eventually offer a permanent and safer form of treatment.

Before we leave hemophilia, it is worthwhile to reflect on a negative consequence of success in treating this X-linked recessive disorder—a *dysgenic effect*. In the past, males with severe hemophilia rarely survived long enough to become fathers; accordingly, the mutant gene was not passed on to the next generation. Now that men with hemophilia can expect near-normal health and life span, their daughters will become carriers of the mutation; if a man with a hemophilia allele mates with a woman who is a carrier, half the female children will be affected, on average. Obviously, this is not an argument for withholding treatment from males with hemophilia, but it does point out that *the need for reproductive planning will increase*. In principle, the same problem (without the sex bias) arises as a result of success in treating any genetic disease that formerly caused early death or such severe morbidity that reproduction would not occur. Incidentally, the fact that most single-gene disorders are rare does not mean that the probability of two mutant allele carriers mating is negligible. The phenomenon known as *assortative mating* applies here: People with a given genetic problem tend to become acquainted with others who have the same problem, one thing leads to another, and suddenly there are offspring with an even bigger problem.

Finally, we must realize that these therapy examples are exceptional situations. There is little hope that nutritional therapies or protein replacement therapies can be devised for most genetic diseases, especially those that affect the central nervous system. However, recent research has opened exciting new possibilities for replacing entire organs or parts of organs, which we will consider next.

Cell and Tissue Replacement Therapies

The previous section outlined some approaches that are being employed to treat genetic disorders one gene at a time. Now we are going to look at the ways in which some human diseases (both genetic and nongenetic) can be treated by replacing entire cells or tissues that are not functioning normally. The oldest of those technologies is *organ transplantation*. Kidneys, for example, fail for many reasons, among which is *polycystic kidney disease*, a genetic disorder that affects at least 1 in 1,000 people worldwide, usually becoming clinically evident in the third or fourth decade of life, when progressive cyst enlargement destroys normal kidney tissue and eventually causes organ failure. Bone marrow transplantation for the treatment of leukemia is another familiar example.

All organ transplantation therapies must cope with immune rejection by the host, although with careful selection of donors (relatives, if possible) and the use of drugs that suppress the immune response, many transplantation recipients have very favorable long-term responses. Another problem is the chronic shortage of suitable organ donors. This has led to attempts to use animal organs for transplantation into humans, but with very limited success, although the possibility of genetically engineering animals to make their organs more compatible with humans is being investigated. Even more speculative, and much in the news recently, is the possibility of growing replacement organs from the recipient's own cells—stem cells.

Stem cells are undifferentiated cells that have the potential to multiply indefinitely and to become one or more types of differentiated cells in response to appropriate external signals. In a sense, the zygote is the ultimate stem cell, but the several hundred cells that constitute *the inner cell mass of the mammalian blastocyst* also are *pluripotent* (capable of becoming many types of differentiated cells), and some of them can be *totipotent* (capable of producing all the cells of the adult, just as the zygote does). Cultures of *embryonic stem cells (ES cells)* can be derived from blastocyts by removing the *trophoblast* (the outer cell layer that is destined to form the placenta and extra-embryonic membranes) and putting the inner cell mass (from which the embryo normally develops) onto a feeder layer of cells that can metabolize but not divide. ES cells from mice have been in use for many years for the production of transgenic animals, which have been particularly useful in analyzing the role of specific genes during development. You will learn more about transgenic mice in Lecture 7.

The procedure for creating embryonic stem cell lines in mice has been extended to several other mammals, including monkeys and humans. The human embryos were available because in vitro fertilization clinics routinely produce more embryos than are needed to achieve pregnancy, so the extra embryos are available for research, when the donors give informed consent and when the research is permitted by law. Announcement of the successful establishment of human stem cell lines in 1998 created great excitement in both the research and medical communities. Research scientists expect to be able to study many aspects of differentiation in unlimited detail with embryonic stem cells, defining the molecular signals that lead to a particular phenotype and identifying all the patterns of gene expression and the proteins that are characteristic of that phenotype. Medical scientists see tremendous possibilities for using stem cells to produce replacement cells for a variety of diseases, such as diabetes (the beta cells of pancreatic islets), Parkinson's disease (dopaminergic neurons), heart disease (cardiac muscle cells), and so on.

However, the immune rejection problem will still need to be solved, even if abundant differentiated cells of a particular type can be generated from stem cell lines. Cells derived from one embryo may be rejected by the host via the usual mechanisms for recognition and destruction of tissue that is non-self. Rejection of foreign tissues is mediated by a highly variable set of cell surface proteins, called HLA antigens in humans, which are encoded by a group of genes, *the MHC complex*, each of which is exceptionally polymorphic. Might it be possible to genetically engineer some human stem cells so that they don't produce HLA antigens at all, and thus could be used as universal donors? It's quite speculative, but that possibility and related ones are being considered.

The foreign tissue rejection problem has also stimulated suggestions that it might be possible to prepare repositories of stem cells derived from each individual, so that when that person needed an organ replaced, the new organ (or at least a rudimentary form of it) could be grown in culture and transferred to the patient. This is really far-out stuff, clearly in the realm of science fiction at present, but there is no theoretical obstacle to achieving it. The basic idea is to take nuclei from somatic cells (maybe stem cells from the bone marrow) soon after birth and inject them into enucleated recipient eggs provided by some cooperative woman (such as the baby's mother), thus creating clones of the baby. When the embryo clones reached the blastocyst stage, which can be done in culture, ES cell lines would be established from them, grown to some convenient amount, and frozen. People with lurid imaginations tend to think that the cloned embryos would be grown to a late fetal stage, cut up into individual organs, and then frozen. Forget it—it isn't possible, let alone acceptable.

The possibility that vast numbers of human embryos might be created by in vitro fertilization specifically for the creation of stem cell lines, rather than as a byproduct of making parenthood possible for couples with reproductive problems, quickly produced a hurricane of media attention. Religious conservatives who believe that a separate person, complete with a soul, is established at conception were adamantly opposed to the use of material derived from human embryos. Liberals pointed out that there were plenty of excess embryos available in freezers at IVF clinics, that those embryos would eventually be destroyed, and that failing to carry out research into the clinical prospects for therapeutic use of stem cells would deny health to many seriously ill persons. Ultimately, President Bush decided that the government would not allow the production of new human embryonic stem cell lines, but research could continue with the existing several dozen lines. Other nations are wrestling with the ethical problems related to human ES lines. It is not yet clear whether a global consensus will emerge.

The stem cell story is not limited to embryonic cells. We have known for a long time that there are *hematopoietic stem cells* (capable of giving rise to all the types of white blood cells as well as erythrocytes) and muscle stem cells. More recently, stem cells have been identified in the fat, brain, and skin. It is quite possible that most organs have a population of stem cells that can provide at least limited regenerative capacity. Moreover, stem cells from one differentiated tissue may be able to differentiate into cells characteristic of another organ. Recent work with mice has shown that bone marrow stem cells can enter the brain and acquire many of the characteristics of neurons. In humans who have received bone marrow transplants, some of the donor cells can be found in the recipient's liver, where they appear to have become hepatocytes.

If it turns out to be possible to produce virtually every type of cell from stem cells that exist in the adult, the problem of immune rejection of non-self tissues will have been largely eliminated. The additional possibility of genetic modification of stem cells while they are in culture suggests that organs that fail because of a genetic defect might be correctable with cells from the same patient, which could then carry out the function that was defective originally. We have a long, long way to go before such therapies enter the clinic, but we are not dealing with anything like warp drives or anti-gravity machines; these biological miracles are not just conceivable, they are possible, given some likely technological advances.

Gene Therapy

The term *gene therapy* refers to the introduction of genes into the human body for the purpose of curing a defect or slowing the progression of a disease. In most cases gene therapy is intended to replace functions that are abnormal because of mutations in a specific genetic locus; it is particularly suitable for treatment of recessive genetic diseases, where the presence of the mutant genes does not interfere with the function of normal copies of the same gene. Gene therapy protocols are also being developed for cancer treatments, where the goal is to introduce genes into the tumor cells that either cause them to commit suicide or to die in response to an external agent (see Lecture 6).

Genes that are artificially transferred into an organism are called *transgenes*. Introducing genes into humans is enormously difficult technically. There must be a way to get the transgene into the target cells, where it must be expressed at appropriate times and in adequate amounts. Additionally, the transgene must not produce adverse effects, such as interfering with the expression of a normal gene or induction of a tumor. Most gene therapy trials make use of modified viruses as vectors for the delivery of transgenes. This takes advantage of basic viral biology, because viruses have evolved to transport their genomes into host cells and take over host cell metabolism to produce virus progeny. A suitable gene transfer vector has to be able to infect the target cells in the host (the patient) efficiently, where the product of the transgene must be produced in amounts sufficient to cure the metabolic defect. It is also essential that no infective viruses be produced in the target cells.

Several types of viruses are being used to develop gene therapy vectors. The most commonly used ones are *retroviruses*, which are RNA viruses that are transcribed into double-stranded DNA in the host cell and subsequently integrated into the host genome. The latter feature is particularly appealing for gene therapy, because it implies that successful integration of a transgene will provide a permanent source of the encoded gene product.

Simple retrovirus genomes contain three genes (Figure 4-4). *Gag* encodes the proteins that bind to the viral genome to form the *nucleoprotein core* of the virus; *pol* encodes *reverse transcriptase* and *integrase*; and *env* encodes proteins that form the *viral envelope*, which is crucial for determining which host cells the virus can enter. Sequences at either end of the structural genes are involved in transcription of the viral genome and binding of proteins (*encapsidation*) to the viral RNA to form viral particles. It is relatively easy to make a gene transfer vector by removing all three protein-coding genes (from the DNA form of the viral genome) and replacing them with a gene (or genes) to be transferred to recipient cells (Figure 4-4). To make a functional gene transfer virus, the modified DNA has to be encapsidated with normal virus proteins, which it can no longer produce. Those proteins can be supplied by co-infection of cells in culture with a helper virus that does have the viral protein genes (but cannot form infectious virus, because of modifications engineered into its genome). Alternatively, the viral proteins may be supplied by cells that have been engineered to produce them constitutively.

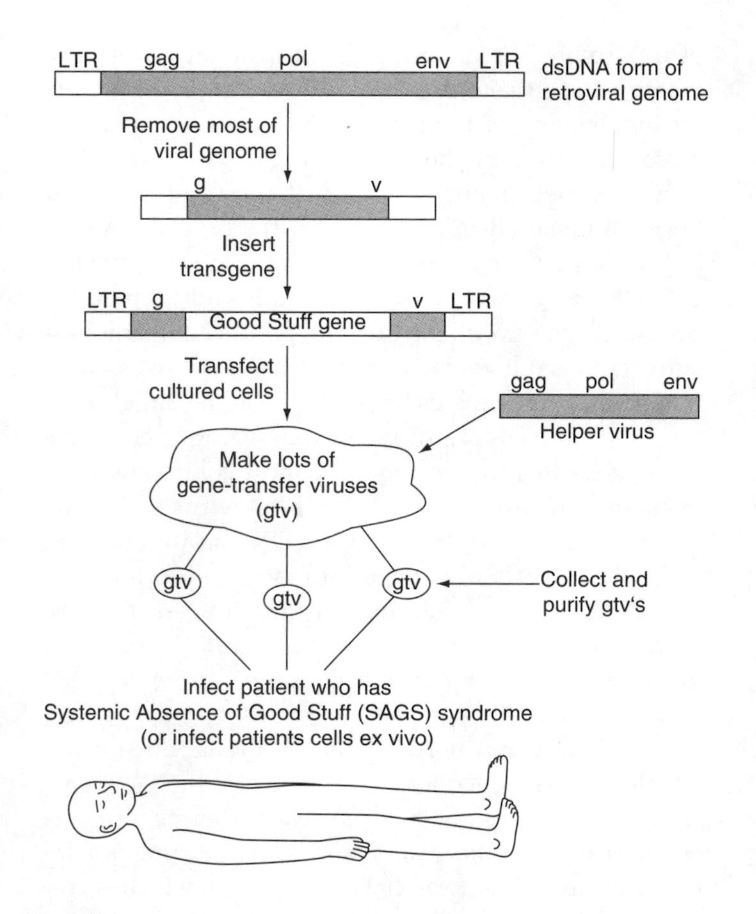

Figure 4-4 Construction and use of gene therapy viruses.

Many gene transfer vectors are based on a simple retrovirus called *Maloney leukemia virus*. Lots of variations have been created to make expression of the transgene more efficient, to make expression of the transgene controllable by an exogenous small molecule, and to decrease the probability that wild-type virus will be formed in the cells used to produce the vector. One disadvantage of simple retroviruses is that they cannot enter the nucleus of nondividing cells. This is a serious problem, because most cells in the adult human body divide rarely, if at all. Thus, the DNA form of the virus cannot be integrated into the recipient cell genomes. Fortunately, there are more complex retroviruses that can enter the nucleus of nondividing cells through the nuclear pores. These are called *lentiviruses*, and HIV belongs to that class. Various efforts to produce gene therapy vectors based on HIV and related viruses are underway. In addition to their ability to integrate into the genome of nonmitotic cells, lentiviral vectors offer the possibility of controlling target cell specificity and transcription of viral DNA in more complex ways than can be done with the simple retroviral vectors. Lentivirus vectors also can accommodate larger transgenes than simple retroviruses.

Gene therapy vectors have also been derived from DNA viruses, principally *adeno-associated viruses* (AAV) and *adenoviruses*. The former are small (4.7 kb) single-stranded DNA viruses with a broad host range, including nondividing cells like neurons, muscle cells, and hepatocytes. For productive infection, they require co-infection with adenovirus or herpesvirus, which provide gene products not included in the AAV genome. In the absence of a helper virus, AAV may integrate into the host genome at a specific site on human chromosome 19, which is desirable for gene therapy, because the transgene can be expressed permanently.

Adenoviruses have double-stranded DNA genomes about 36 kb long. They can infect many types of human cells, including the liver, gastrointestinal tract, and respiratory epithelium. The adenovirus genome enters the nucleus of infected cells, where it replicates and is transcribed, but does not integrate into the host genome. Although an adenovirus infection may persist for weeks, the ultimate fate of infected cells is lysis and death. A major advantage is that substantial portions of the adenovirus genome can be deleted, so that a relatively large transgene can be incorporated into the gene transfer vector. As was the case with retroviruses, the deleted genes that are required for packaging of vector DNA into mature virus particles have to be supplied in *trans* by concurrent transfection with a helper virus or by the use of cells that stably produce the necessary proteins.

A significant problem with the use of gene transfer vectors based on adenovirus is the powerful immunogenicity of some of the viral proteins, which can pose a significant risk to patients who may have already been sensitized by previous infections. That fact, plus the normal property of adenovirus infections to lyse host cells, can be exploited to design cancer-killing agents. Special strains of adenovirus have been constructed that replicate preferentially in tumor cells. In addition to their direct *cytocidal effect*, the special adenoviruses elicit an anti-tumor response by the host immune system. It's a double-whammy strategy, still in early stages of development, but very promising.

What clinically significant accomplishments have been achieved with viral gene therapy vectors? So far, success has been limited, but the problems inherent in gene therapy are so complex that we should not be surprised if it takes many years to find out what is possible and what is not. Considerable media attention has focused on several attempts to cure *severe combined immune deficiency (SCID)*, beginning in 1990 with a girl whose immune system was nonfunctional because she had two mutant copies of the gene for *adenosine deaminase (ADA)*. This enzyme is widely distributed in the body, but its function of removing certain metabolic waste products is particularly crucial for thymus-derived lymphocytes (T cells), in two ways. First, thymus differentiation is abnormal if insufficient ADA is present; thus, T cells are not produced in normal amounts. Second, the T cells that are produced have abnormally brief lifetimes. Without T cells the entire immune system fails.

Fortunately, it is possible to purify ADA from other sources and attach it to a polymer (*polyethylene glycol, PEG*) that stabilizes it so that it is not quickly degraded by enzymes in the blood. SCID patients who have ADA deficiency can be kept alive that way, although at great expense and with constant care. Ashi, the gene therapy pioneer patient in 1990, was maintained on ADA-PEG. For gene therapy, a large number of T cells were collected from her blood, transfected with a retrovirus that carried the human ADA gene, then transfused back into her blood. Although some of the engineered T cells did synthesize ADA, it was not enough to protect Ashi from infection, so she has remained on ADA-PEG therapy. Currently, doctors are treating another SCID patient by isolating stem cells from her bone marrow, transfecting them with an ADA-transfer virus, and then returning them to her blood by transfusion. The results are not yet known, but in principle, any beneficial effect should be permanent.

Nonviral methods for delivering genes are also under active investigation. The simplest method seems to be *injection of naked plasmid DNA* into muscle. A significant fraction of muscle cells take up the DNA and produce the encoded protein, at least for several weeks. An improvement on direct injection involves clamping an artery that feeds a muscle for a few seconds, so that hydrostatic pressure builds up in the artery, then injecting DNA into that artery. Apparently the high pressure forces the DNA out of the artery and into adjacent cells. Other researchers are testing a technique called *electroporation*. A brief electric shock applied to target tissue such as skin or muscle punches holes in the cell membranes transiently and allows injected DNA to enter. There are also numerous trials of techniques that wrap DNA in lipids or other polymers to make it more likely to be taken up by target cells.

At least 400 clinical trials of gene therapy have been done or are in progress worldwide. Most of them are aimed at controlling or curing cancer and are conducted on terminally ill patients (see Lecture 6). Other gene therapy protocols intended to cure AIDS are in progress. Relatively little effort is focused on diseases like SCID, because of their rarity. Indeed, the tongue-in-cheek comment that the number of researchers exceeds the number of SCID patients is not far wrong. For more common recessive genetic diseases (CF, PKU, sickle cell disease, muscular dystrophy, etc) there have been some encouraging experimental results, but so far nothing close to a complete and permanent cure has been achieved.

What should we expect from gene therapy ultimately? Disorders of the hematopoetic system will be the most treatable, because the cells in peripheral blood can be manipulated *ex vivo*, then returned to the patient. The fact that hematopoietic stem cells can be isolated and genetically engineered (in principle) makes the prospects for successful gene therapy of defects that produce abnormal red blood cells or defective immune functions even more promising. What else can we look forward to? Will every baby born with cystic fibrosis or PKU receive an injection of a "magic bullet" gene transfer vector immediately after birth and be able to function normally thereafter? Well, maybe so; it's not a totally naïve expectation, but we are a long, long way from fulfilling it. Neurological diseases of genetic origin will be more problematic, because of the difficulty of getting access to the affected cells. Gene therapy for dominant genetic disorders of the dominant negative and gain-of-function types (see Lecture 3) will require inactivation or removal of the mutant allele, rather than simply providing a normal allele, which is a serious technical problem. For now, cautious optimism about the future of gene therapy is reasonable, but frothy enthusiasm is not justifiable.

Pharmacogenomics

We are not all alike metabolically. One of the important ways in which we differ is our response to drugs. Your post-operative pain may not be relieved as much by codeine as my pain; you may feel that Prozac makes life more wonderful than it has ever been, while your "significant other" mostly responds to the drug with diminished libido; a specific type of cancer chemotherapy may be very effective in some patients, marginally effective in others, and toxic in a few others. All medications have variable effects on the people who take them. Much of this variability is related to a patient's age, general health, lifestyle, and other environmental factors, but it is becoming increasingly clear that subtle variations in genotype have a profound influence on responses to medication.

Recognition of the effects of genes on drug efficacy and safety led first to a field of study called pharmacogenetics, which dealt with inherited variations in single genes, then to pharmacogenomics, which uses more modern techniques to study the influence of the entire genome (or large groups of genes) on drug responses. Because most drug responses can be influenced by several to many genes, pharmacogenomics is the term most frequently used now.

How can genetic variations influence a person's response to a particular drug? Let's think about that question at two levels: *biochemical* and *genetic*. At the biochemical level variations in enzyme activity and/or enzyme specificity can affect *drug metabolism*; that is, inactivation of an active form of a drug, activation of an inactive drug, or transport of a drug into its target cells. A second type of biochemical effect is variations in the *affinity of drug receptors*, the molecules to which a drug must bind in order to exert its desired metabolic effects.

At the genetic level, we must realize that interindividual variations in drug responses are not merely the presence or absence of an enzyme or drug receptor. Undoubtedly, most variability in drug responses is related to polymorphisms that have quantitative effects. It would be simplistic to expect people to fall into two distinct classes: responders and nonresponders to a given drug. Instead, the numerous proteins with which a drug interacts directly and the even more numerous proteins that the drug may affect indirectly guarantee that the patient population will exhibit a spectrum of responses.

Here are a few examples:

1. A protein called *P-glycoprotein (PGP)*, the product of the *multidrug resistance gene (MDR-1)*, interacts with a substantial fraction of all drugs used in humans. It functions as a pump, expelling foreign substances from inside of cells to the outside. In persons with polymorphisms in MDR-1 that lead to overexpression of PGP, higher concentrations of drugs are needed to achieve the desired effect.
2. One of the P450 family of detoxifying enzymes is the product of the *CYP2D6* gene, and one of its properties is conversion of opioid analgesics to an active form. Several percent of the human population are homozygous for nonfunctional *CYP2D6* alleles and therefore receive little relief from codeine. The CYP2D6 protein also metabolizes *fluoxetine* (a common anti-depressant); standard doses of that drug can be fatal to CYP2D6-deficient patients.
3. Thiopurines of several types are used in treatment of cancers. In persons deficient for the enzyme TPMT, which inactivates thiopurines, standard doses can produce serious hematopoietic toxicity.

How are the genetic variations that influence drug responses identified? This requires the interaction of clinical and genomic information. The initial observation that people do vary in their response to a particular drug can be followed by a search for variations in the genes that encode the proteins with which a drug interacts; that is, the enzymes that change the drug, the molecules involved in transport of the drug, and the proteins that the drug affects directly. This process becomes easier as our knowledge of human biochemistry increases, along with more detailed understanding of our genome. However, the task becomes more difficult when significant variations in drug responses are caused by polymorphisms in the genes coding for proteins downstream of the drug's direct effects. A systems analysis approach becomes essential for understanding those downstream effects, which is now feasible with microarray analysis.

Pharmaceutical and biotech companies are now actively looking for correlations between SNPs and variations in drug response. The effort is massive, and will surely yield useful results in the near future. A related area of research is *rational drug design*, which uses detailed information about the structure of proteins to design drugs that are optimized to fit into a catalytic or regulatory site. A prominent example of rational drug design was the development of specific inhibitors of cyclo-oxygenase 2 (COX2), whose active site differs from the active site of COX1 by only one amino acid. COX2 inhibitors are widely used to treat arthritis, without the gastric damage that often accompanies the use of nonspecific anti-inflammatory drugs. Eventually, combining our knowledge of SNPs, protein structure, and drug responses with detailed information about each patient's genotype should lead to individually optimized treatment for many diseases, both genetic and nongenetic.

BOX 4-1
MICROARRAY ANALYSIS OF GENES AND GENE EXPRESSION

Microarray analysis is an immensely powerful technology that has become available in the last few years. It is revolutionizing the analysis of gene expression, because it allows an investigator to measure the level of thousands of mRNAs on a single slide. Rather than asking whether the expression of gene ABC changes in response to stimulus X, one can find out almost *everything* that happens at the mRNA level in response to an experimental variable or a disease or natural variation. Patterns of response involving dozens or hundreds of gene products are becoming evident from the application of microarray analysis.

The diagram outlines preparation of microarrays, which can be done several ways. Each DNA spot on an array is quite tiny; 30,000 or more spots may be put onto a treated slide roughly 1 inch square. Arrays are then used to probe a test sample and a reference sample (usually labeled with fluorescent dyes), which can hybridize to complementary sequences in the DNA on the arrays. Software has been developed to allow extensive quantitative and comparative analysis of the results.

In the context of this lecture, you can imagine that the array contains oligonucleotides representing places in the genome where SNPs are known to occur. Then, hybridization of PCR-amplified segments of an individual genome can accomplish a genome-wide scan for SNPs in one assay. This has applications to designing individualized therapeutic regimens and to mapping genes associated with complex diseases (Lecture 3). A common use for microarrays is to hybridize a set of cDNAs from a particular cell type to DNAs that represent most or all of the genome. This application of microarrays is proving to be very helpful in defining subtypes of cancer (Lecture 6); microarrays also have exciting potential for the study of gene expression during development, both normal and abnormal.

Microarray Analysis of Genes and Gene Expression

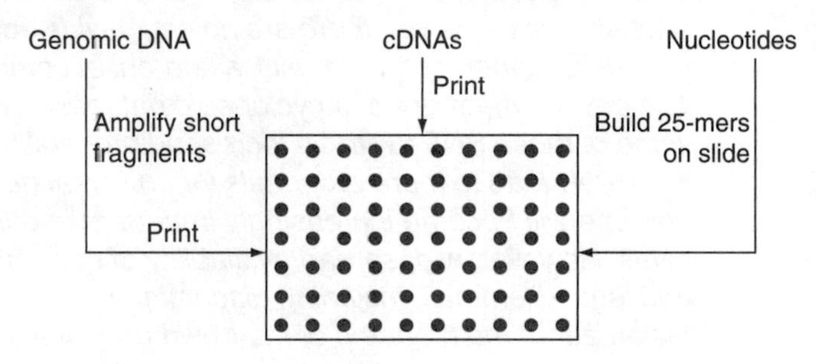

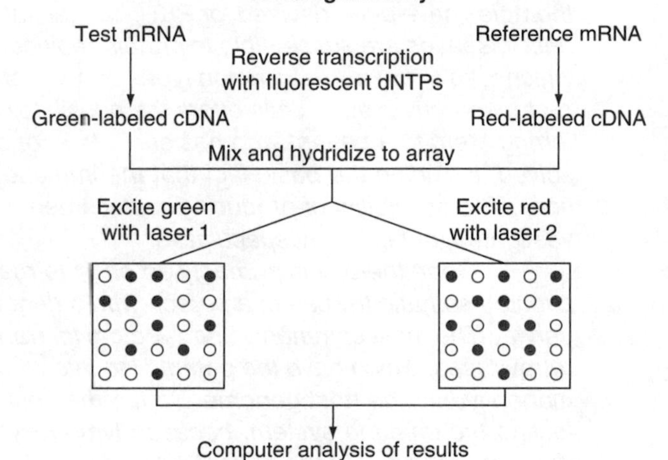

REVIEW

This lecture summarized the impact of human genome research on medical practice. Genetic screening of newborns began with programs to identify infants with phenylketonuria and has expanded to cover several other easily detectable metabolic abnormalities. Adult screening is currently focused on identifying couples at risk for the birth of a child with a genetic disease, such as cystic fibrosis or Tay-Sachs disease. As yet, there are no nationwide adult screening programs, although proposals to screen the public for hemochromatosis, a late-onset common defect in iron metabolism, have been considered. The development of an encyclopedic database on single-nucleotide polymorphisms and correlation of some of those SNPs with an increased probability of developing one of the common complex diseases will surely lead to more proposals for universal genetic screening, because the details of an individual's genome will become increasingly important in offering both preventative and therapeutic advice to patients. However, widespread availability of genomic information about individuals leads to many ethical and legal dilemmas, including discriminatory actions by insurers or employers, as well as patients' confusion about the meaning of their own genotypes.

Many types of therapy for genetic disorders are available. Some highly successful nutritional therapies have been devised for PKU, congenital hypothyroidism, and several other disorders. A few genetic diseases are susceptible to protein replacement therapies. These include defects in blood clotting, which lead to the more common types of hemophilia and some of the lysosomal storage diseases. Current research on stem cells offers the possibility of growing many types of tissues in culture and transferring them to a patient who has some type of organ failure. Serious technical problems remain to be solved, including the basic fact that the immune system will usually reject transplants from other persons. The possibility of producing replacement tissues from an individual's own stem cells is being investigated, but is highly speculative.

Gene therapy is primarily an effort to restore normal function by introducing normal genes into affected somatic tissues of a person with a genetic disease. Hundreds of trials of gene therapy are underway. The most commonly used vectors for transferring genes to target tissues are specially designed retroviruses, which have the potential to infect target cells and ultimately insert a functioning gene permanently into the host genome. So far, the best results have been obtained with genetic diseases affecting the immune system, because lymphocytes can be obtained from peripheral blood, altered ex vivo, *and returned to the patient.*

Recognition of the fact that variations in drug response often have a genetic basis, combined with increased knowledge of those genetic variations, has led to the rapidly expanding field of pharmacogenomics, which seeks to use genomic information to design optimized therapeutic regimens for individuals. Overall, it is clear that the more we understand the structure, variation, and function of human genes, the more powerful and effective medical practice will become.

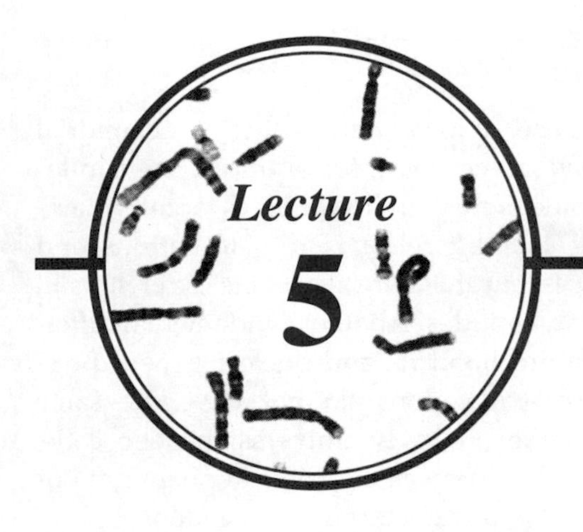

The Mitochondrial Genome and Mitochondrial Disorders

The preceding four lectures were focused on nuclear genes, which constitute greater than 99% of human genes. Now it is time to learn about "the other genome," the DNA contained in mitochondria. In humans, as in other mammals, the mitochondrial genome is 16.6 kb in size, just enough DNA to encode 13 polypeptides and two dozen RNAs. Despite its small size, the mitochondrial genome is absolutely essential for normal metabolism in nearly all cells, and the many ways in which mitochondrial functions can be perturbed constitute a significant fraction of human genetic disease. Before we discuss genetic diseases, let's begin with a brief summary of what mitochondria are and what they do.

Mitochondria are complex organelles—the evolutionary descendants of prokarytotes that became intracellular symbionts of primitive eukaryotes in eons past, before plants and animals diverged from a common ancestor. There are significant differences in the structure, biochemistry, and genetics of mitochondria among major eukaryotic groups (plants, fungi, protists, and animals), and it may be that several endosymbiotic relationships evolved independently into modern mitochondria. Mitochondria range widely in size; they may be ellipsoidal, as in mammalian liver (roughly 1 by 2 micrometers); or they may have a cylindrical form, about 1.0 micrometer in diameter and several to many micrometers in length. Their form is not static; mitochondria grow, bud, break off pieces, fuse with other mitochondria, and separate again. In fact, the mitochondria in some cells can be interconnected to form a reticulum, at least under certain growth conditions. Most mammalian cells contain several hundred mitochondria; some contain many more. Red blood cells have none. *The major structural features are the two membranes and the cristae*, which are convolutions of the inner membrane that project into the *matrix*, or interior of the organelle (Figure 5-1).

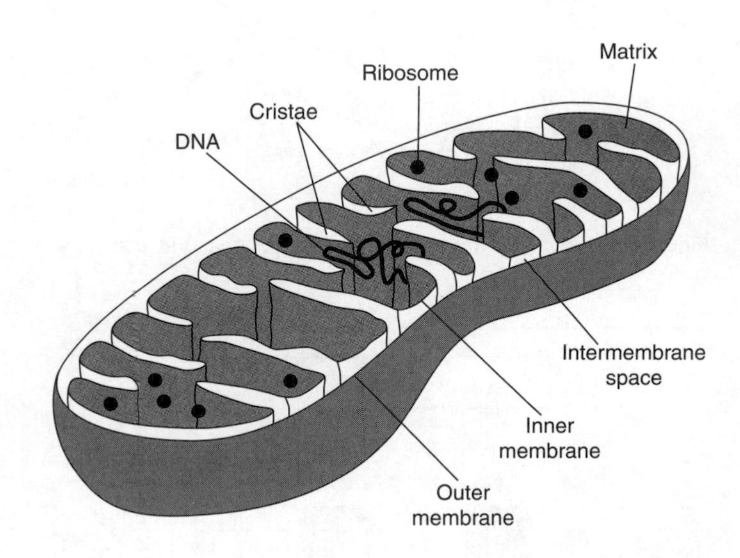

Figure 5-1 Mitochondrial structure.

Mitochondria are the principal sites of cellular energy production. Most of the ATP generated in eukaryotic cells is formed in mitochondria, via a series of reactions termed *oxidative phosphorylation*. The high-energy electron carriers, NADH and FADH2, from which the energy for ATP synthesis is indirectly derived, are produced by the catabolism of carbohydrates, lipids, and proteins (Figure 5-2) *Acetyl CoA* is produced within the mitochondria from *pyruvate* (the product of glycolysis), from *fatty acids* (the product of lipid catabolism) and from *amino acids* (the product of protein catabolism); then acetyl CoA is oxidized to CO2 and water via the *citric acid cycle*, producing the electron carriers NADH and FADH2 concomitantly. The latter compounds enter the *respiratory chain* (Figure 5-3), where their available electrons are transferred stepwise to oxygen, with the simultaneous export of protons (hydrogen ions) through the inner mitochondrial membrane to the inter-membrane space. The *chemiosmotic gradient* thus created is the source of energy to make ATP from ADP and inorganic phosphate as hydrogen ions pass through the protein complex known as *ATP synthase* back into the mitochondrial matrix.

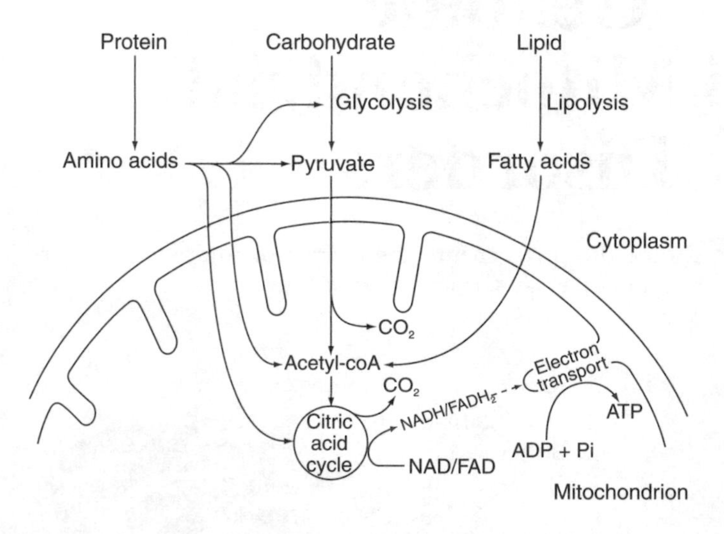

Figure 5-2 Sources of metabolites for mitochondrial oxidation.

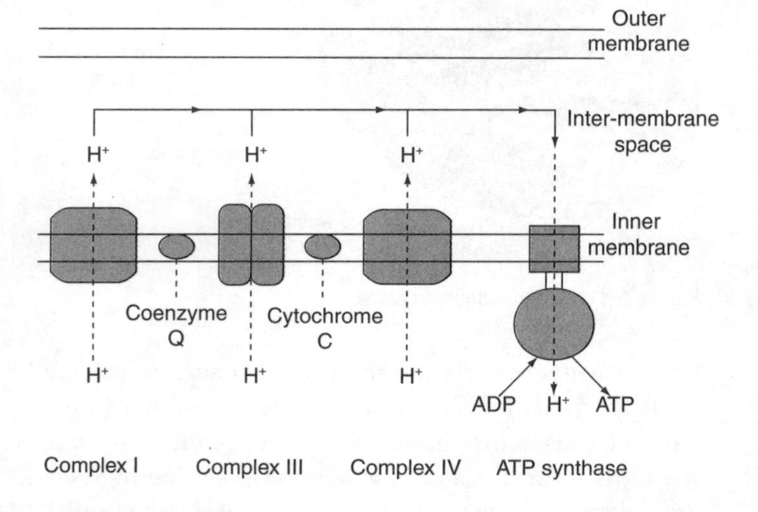

Figure 5-3 The major steps in oxidative phosphorylation. Note that the components are not physically organized as shown; they move around freely within the membrane.

The 16.6 kb of human mitochondrial DNA (mtDNA) encodes 13 polypeptides of the respiratory chain, 2 ribosomal RNAs, and 22 tRNAs (Figure 5-4) Mitochondrial DNA is noteworthy for its density of information; there are no introns, and in several instances the first nucleotide of one gene is also the last nucleotide of the preceding gene. In lower eukaryotes, mitochondrial DNAs are much larger; yeast, for example, has a mitochondrial genome more than five times as large as mammals. Most of the difference is due to introns and intergenic spaces, but there are also some differences in the encoded polypeptides. Another fundamental difference between mitochondrial and nuclear gene expression is that several codons do not have the same meaning as in the otherwise-universal genetic code; UGA, for example, is translated as tryptophan in mitochondria, instead of being a termination codon.

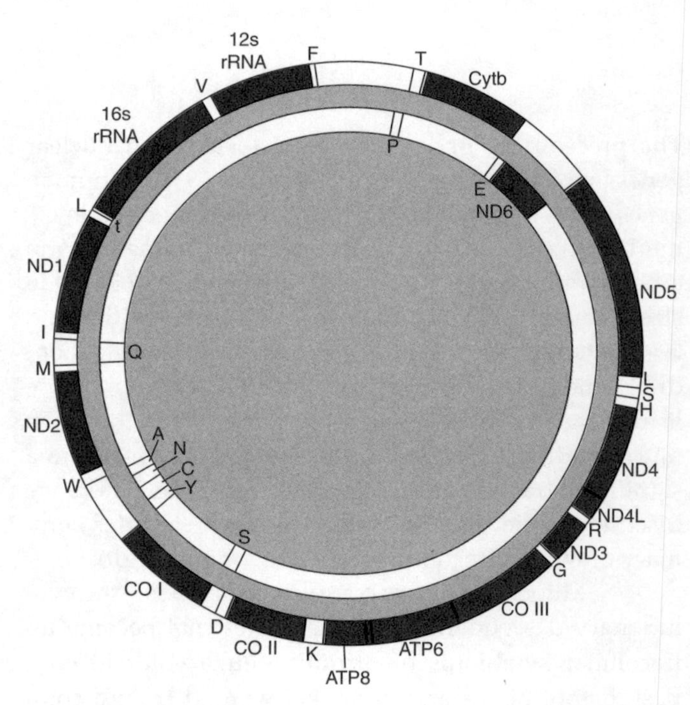

Figure 5-4 The human mitochondrial genome. Each circle represents one strand of DNA. Dark shaded regions are the rRNA and polypeptide genes. Symbols are: ND1–6 encode subunits of NADH dehydrogenase (complex I); Cytb is cytochrome b (part of complex III); CO I, II, and III are cytochrome oxidase subunits (complex IV); ATP6 and ATP8 are subunits of ATP synthase. The tRNA genes are indicated by light blocks labeled according to the amino acid the tRNA carries. From Wallace, D. C., 1989. *Cytogenetics and Cell Genetics 51*, p613. By Permission of S. Karger Inc.

DNA synthesis, RNA synthesis, and RNA processing within mitochondria all depend upon enzymes encoded by nuclear genes. Similarly, all of the protein components of mitochondrial ribosomes and the various accessory factors needed for protein synthesis are the products of nuclear genes in mammals; but they are all different from the genes that code for the cytoplasmic protein synthesis machinery. In all, hundreds of proteins are involved in mitochondrial structure and contents, and nearly all of them are encoded by nuclear genes. Getting those proteins from their sites of synthesis in the cytosol into their proper places within mitochondrial membranes or the mitchondrial matrix requires some special transport mechanisms.

The *outer mitochondrial membrane* is freely permeable to small molecules, up to a mass of 6,000 daltons, which can pass through tiny pores created by the membrane protein, *porin*. The *inner mitochondrial membrane* is impermeable to all but the smallest uncharged molecules (e.g., O_2 and H_2O); an unusual lipid called *cardiolipin* is thought to make a significant contribution to that impermeability. This creates a problem for mitochondrial function and mitochondrial assembly. First, the substrates on which the oxidative enzymes in the matrix act must be imported via a group of *translocases*—proteins located in the inner mitochondrial membrane; for example, there are specific translocases for pyruvate and for ADP. The translocases themselves, the oxidative enzymes within the matrix, the nucleic acid and protein synthesis apparatus, and many other components of the mitochondrial membranes are all synthesized on cytoplasmic ribosomes. They all contain a short sequence of amino acids at the beginning of the protein that targets them to mitochondria. In addition, many chaperone proteins and assembly proteins (not well known) help mitochondrial proteins pass through or into the two mitochondrial membranes; these are proteins that do not remain in the mitochondria, but they are essential for mitochondrial assembly and function. The major components of the respiratory chain are complex proteins containing both nucleus-encoded and mitochondrial-encoded polypeptides. These relationships are summarized in Figure 5-5.

The endosymbiont hypothesis of mitochondrial origin implies that they originally contained a complete cellular genome. For reasons that we don't fully understand, most of that "captured" genome has been either lost or transferred to the nucleus of the host cell, where some genes have undergone further evolution. Two mechanisms for transfer have been postulated. One proposes that occasionally a cDNA copy of a mitochondrial mRNA is made and integrated into the nucleus, by analogy with the origin of processed pseudogenes (mentioned in Lecture 1). However, this model requires that reverse transcriptase has access to mitochondrial mRNAs. It is hard to see how that could happen in intact mitochondria, but if mitochondria occasionally lyse, liberating their contents into the cytoplasm, then it would be possible for cDNAs to be made from mitochondrial mRNAs. The other mechanism, direct transfer

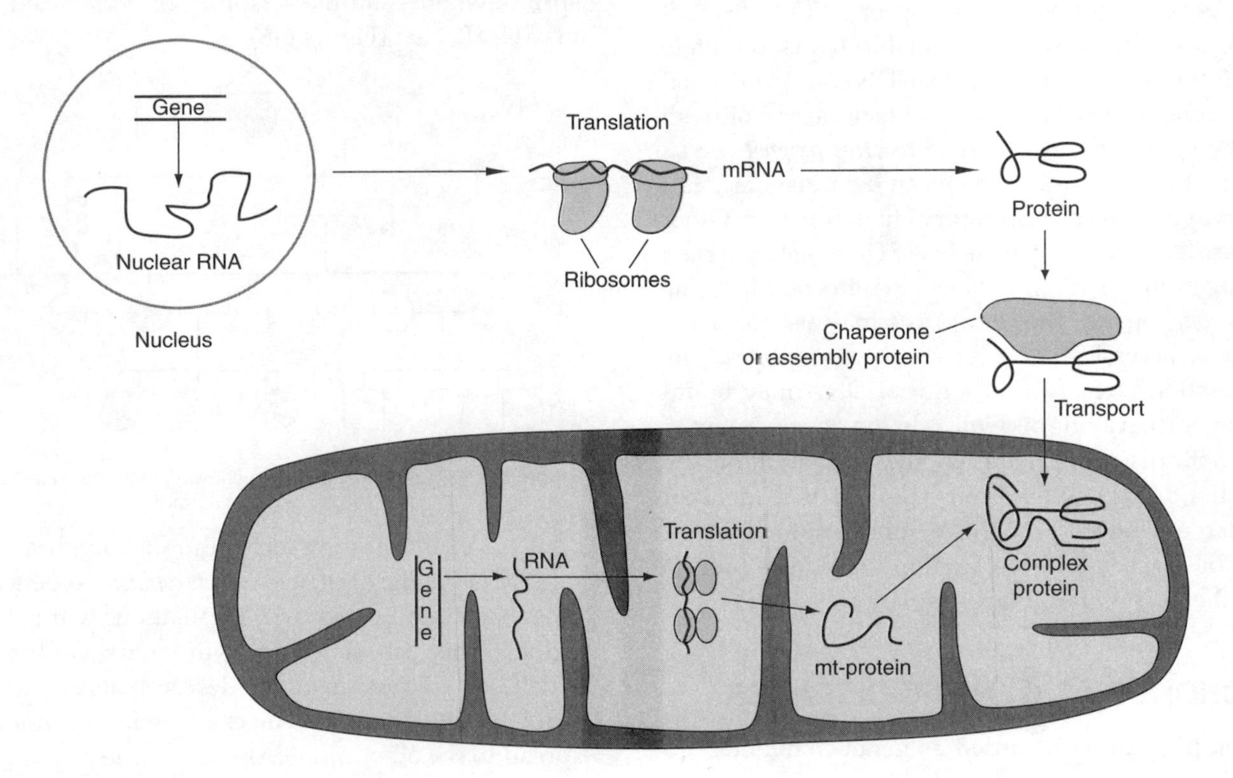

Figure 5-5 Synthesis and assembly of complex mitochondrial proteins.

of mitochondrial genes to the nucleus, also postulates mitochondrial lysis, followed by incorporation of one or more mitochondrial genes into the nuclear genome. Possibly some sort of injury to a cell would cause both mitochondrial lysis and nuclear DNA damage, which would offer an opportunity for a fragment of mitochondrial DNA to be spliced into a nuclear chromosome. It's unlikely, of course, but it doesn't have to happen often—perhaps once every million years.

Why haven't the genes that remain in mtDNA been transferred to the nucleus? We can only speculate that it may be exceptionally difficult for mechanisms for their gene products to be transported from the cytoplasm into the mitochondria to arise. This is easy to visualize for the RNAs with their high density of negative charges from the phosphate groups; for such molecules to penetrate the mitochondrial membranes would require some complicated carrier system. The very hydrophobic proteins that are still encoded by the mammalian mitochondrial genome may present special transport and entry problems, or the differences in the genetic code may be an insurmountable barrier to transfer. Perhaps in another billion years evolution will have solved those problems. Stay tuned.

The existence of a separate genome within a cytoplasmic organelle raises the possibility that mutations in those genes might have special properties. Indeed, it is well known that mitochondrial DNA tends to mutate about 10 times faster than nuclear DNA. It is believed that the difference is due to several factors, one of them being the constant generation of *reactive oxygen species* (mentioned in Lecture 2), which can be mutagenic, as a common byproduct of respiratory chain function. Other factors are likely to be *lower fidelity of DNA replication and less accurate DNA repair mechanisms* in mitochondria. The *lack of histones* on mitochondrial DNA may also make the DNA more accessible to reactive oxygen species. The high mutation rate of mitochondrial DNA may be involved in normal aging, as well as in the development of overt genetic disease. In addition, the high mutation rate of mitochondrial DNA has been exploited by students of molecular evolution, who have found mitochondrial DNA to be a rich source of variation that can be used to deduce the history of human populations.

MITOCHONDRIAL GENETIC DISEASES

Mitochondria can be regarded as genetic chimeras: 13 essential polypeptides are encoded by the mitochondrial genome but hundreds of other polypeptides that make up a mitochondrion are encoded by nuclear genes. Mutations in nuclear genes will generally behave in the classic Mendelian fashion, causing diseases that are mostly inherited either in the standard autosomal recessive or autosomal dominant patterns. But what about mutations in the mitochondrial genome? Mitochondria don't undergo mitosis; an average mitochondrion contains 5 to 10 mitochondrial DNAs, which don't replicate in synchrony with nuclear chromosomes, and mitochondria tend to be randomly distributed to daughter cells when cells divide. These facts are enough to make mitochondrial inheritance complicated, but there's more to the story.

Diseases Attributable to Mutations in mtDNA

The definitive characteristic of genetic diseases caused by mutations in mtDNA is *maternal inheritance*. Although sperm contain mitochondria and depend upon them for the energy needed to move up the female genital tract, those mitochondria are in the mid-piece, which does not accompany the sperm nucleus into the egg that it fertilizes. Therefore, all mitochondria in an embryo are derived from the ovum. A male affected with a mitochondrial disease cannot pass the disease to his offspring, whereas all the offspring of an affected female may be affected (Figure 5-6).

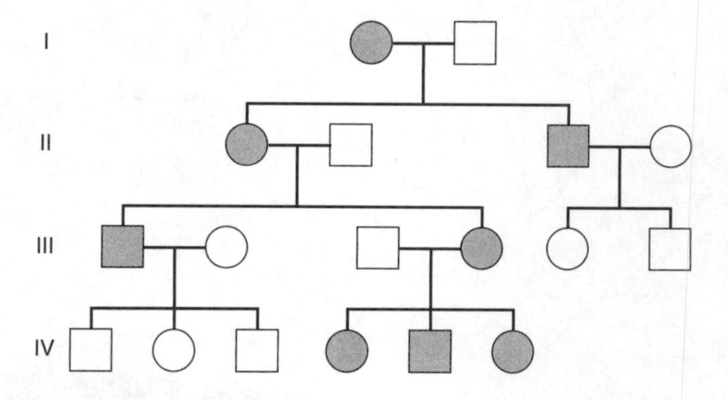

Figure 5-6 Hypothetical pedigree showing maternal inheritance.

A complicating factor in mitochondrial disease inheritance is the existence of *heteroplasmy*—genetic heterogeneity of mitochondrial populations within an individual. When a mutation arises in a mitochondrial DNA, it will be replicated in all the descendants of that particular DNA, but there is no mechanism for the mutation to spread to the other mtDNAs in the same cell. Nevertheless, when a cell containing a mixture of normal and mutant mtDNAs divides, the daughter cells receive a random mixture of the mitochondria. Thus, the level of heteroplasmy in each new cell may be greater or less than it was in the parental cell. Because most mutations in mtDNA have a negative effect on energy metabolism, it is easy to see that if all the mitochondrial DNA in a given cell were mutant (*homoplasmy*), the cell probably would not be viable. We don't have any definite rule for the amount of heteroplasmy that can be tolerated before a person becomes clinically affected; it varies greatly with the nature of the mutation and apparently also with the genetic background in each individual.

An important clinical question is whether the children of a woman with a mitochondrial genetic disease will be more or less severely affected than their mother. There are about 100,000 mtDNA molecules in a human ovum. Some investigators have suggested a "genetic bottleneck" during oogenesis—a time when the number of mitochondria per cell is very small. If true, then random partitioning of mitochondria could lead to gross differences in the level of heteroplasmy in each ovum. However, that hypothesis is unproven. Another factor that may alter the level of heteroplasmy in children of an affected mother is the fact that mitochondrial DNA does not replicate until after the human embryo implants into the uterus. The 100,000 mtDNA molecules that are present in the ovum are randomly distributed to cells as the blastocyst forms, so it is quite possible that various levels of heteroplasmy may be present in the blastocyst cells. The descendants of one blastocyst cell can represent a substantial fraction of an adult organ or tissue. Moreover, the fraction of mutant mitochondrial DNA molecules may change differentially within tissues as the individual develops, matures, and ages. The result is that some mitochondrial diseases have highly variable symptoms, not only from family to family, but within families. Thus, the clear-cut pattern of maternal inheritance exemplified by Figure 5-6 may not occur in all pedigrees where a disease caused by a mtDNA mutation exists, and that possibility is shown below.

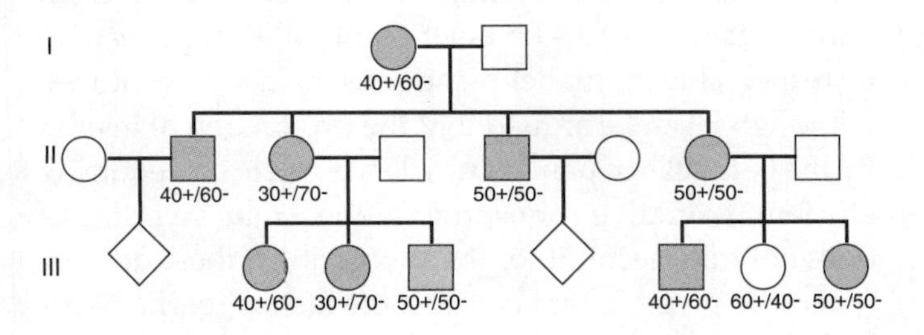

Figure 5-7 Hypothetical pedigree to illustrate heteroplasmy and random distribution of mtDNA at mitosis. The numbers below each symbol give the percent normal mtDNA (indicated by +) vs. mutant mtDNA (indicated by -) in that individual. Note the unaffected female at the right side of generation III; by chance, she received too few mutant mtDNAs to develop an abnormal phenotype.

Another way to categorize genetic diseases caused by mtDNA mutations is to divide them into those that are caused by mutations in one of the 13 protein-coding genes versus those that are caused by mutations in the rRNA or tRNA genes. Mutations in any of the RNA genes in mtDNA will affect protein synthesis generally, thus reducing the number of functional copies of all the proteins encoded by mtDNA. Thus, mutations in mtRNA genes are unlikely to reach homoplasmy, because the overall reduction in energy metabolism would make cells nonviable, unless the defect was very mild. Now we will look at some examples of diseases that arise because of mutations that affect mitochondrial functions.

Myoclonic epilepsy and ragged red fiber disease (MERRF) is caused by an A-to-G transition in the gene for mitochondrial lysine-tRNA. MERRF is characterized by central nervous system abnormalities (e.g., epilepsy, deafness, dementia) and deficiencies of skeletal and cardiac muscle function, associated with abnormal appearance of mitochondria in the electron microscope and "ragged" muscle fibers after staining with trichrome Gomori stain. Severity and type of symptoms vary substantially within a pedigree, which implies heteroplasmy of the mutant mitochondrial DNAs.

Biochemical analysis on muscle biopsies from some MERRF patients showed that there was deficient function of complexes I and IV. No mutations, other than the A to G transition, correlated reliably with MERRF symptoms in those affected lineages. It is an easy mutation to assay, because it creates a cleavage site for a restriction enzyme. Using that enzyme, researchers have shown a good correlation between the severity of symptoms and the proportion of mutant DNA in an individual's mitochondria.

Why does this mutation affect complexes I and IV predominantly, rather than all components of the respiratory chain? We don't know for sure, but basic molecular biology suggests some explanations. First, the mutant tRNA may not be completely inactive, and since the frequency of lysines varies from protein to protein, the amount of complete protein synthesized may vary. Second, the effect of a reduced amount of one polypeptide may have different effects on the stability of an entire complex. Perhaps you can think of other possibilities.

Deafness can also be a mitochondrial genetic disease. It has been known for many years that people treated for bacterial infections with *aminoglycoside antibiotics* (such as streptomycin or gentamicin) sometimes lose their hearing. The explanation for this result lies in the evolutionary history of mitochondria, which (as you know) are descendants of bacteria. Aminoglycosides kill bacteria because they interact with their ribosomes, stabilizing the binding of mismatched aminoacyl-tRNAs, thus leading to the accumulation of errors in proteins. In most humans, mitochondrial ribosomes are sufficiently different from bacterial ribosomes that aminoglycoside antibiotics do not disrupt mitochondrial protein synthesis significantly. However, several families have been identified with a single A-to-G mutation at nucleotide 1553 of the small mitochondrial rRNA, which makes them sensitive to aminoglycosides. In at least some families with that mutation, it is homoplasmic.

Apparently, failure of mitochondrial protein synthesis in the cochlea leads to the death of nerve cells that transmit sound signals to the brain. A fascinating question is why antibiotic treatment has such a specific effect in susceptible individuals. Why don't nerve cells throughout the body die or at least malfunction? We are again reminded that the control of gene expression can vary greatly from one cell type to another. Further evidence of complexity is provided by a five-generation Arab family with the same 1553 A-to-G mutation as the families mentioned above. In this Arab family, which has apparently had no exposure to antibiotics, maternally inherited deafness occurs spontaneously in many, but not all, family members. Genetic analysis has implicated a nuclear gene that modifies the effects of the rRNA mutation. This is only one of several lines of evidence indicating that mitochondrial genetic diseases cannot all be categorized as showing either maternal or simple Mendelian inheritance; complex inheritance patterns also occur.

An example of a genetic disease caused by a *missense mutation* in a protein encoded by mtDNA is *Leber's hereditary optic neuropathy* (LHON). The condition is characterized by a loss of vision, beginning centrally and proceeding peripherally. It is the result of decline of optic nerve function, usually beginning at ages 20–24. Cardiac functions may also be abnormal. Missense mutations in a subunit of complex I account for nearly all cases. It is not known why the optic nerve is the primary locus where this mutation has its effect, nor why it takes two decades for deterioration of vision to become apparent. There is some evidence that nuclear genes also contribute to the probability of LHON symptoms developing in persons with a missense mutation in complex I.

Several mitochondrial genetic diseases are the result of *large deletions* in mtDNA. The age of onset varies, but there is a general tendency for symptoms to become worse over time, often leading to death from respiratory failure or general systemic malfunction. An example is *Kearns-Sayre syndrome,* which is characterized by a variety of neuromuscular symptoms, including retinal degeneration, heart disturbances, dementia, seizures, and others. Numerous patients with Kearns-Sayre syndrome have been shown to have large deletions in a substantial fraction of their mitochondrial DNA molecules. The deletions all involve several mitochondrial genes, but the only consistent feature is that the origins of DNA replication are not deleted (see Figure 5-4). Heteroplasmy is a necessary feature of these deletions; otherwise, cell death would ensue from total failure of essential mitochondrial functions.

Within an individual who has Kearns-Sayre syndrome, all deleted DNAs have the same deletion, which implies that all are descended from a single deletion event. In nondividing tissues (such as nerves and muscles), the proportion of mtDNAs that contain the deletion increases with time. This is puzzling, but one explanation may be that the smaller deletion-bearing molecules take less time to replicate. Another hypothesis is that nuclei can initiate replication of nearby mitochondria in response to a local energy deficit. This suggests that when a cluster of deletion-bearing mitochondria happens to develop near the cell nucleus, their replication is preferentially stimulated, thus making a bad situation worse.

In Lecture 3, we discussed diabetes as an example of a disease with complex causation. Several mtDNA mutations have been shown to be associated with maturity-onset diabetes. One is an A-to-G transition in a gene for leucine-tRNA, which has the effect of impairing transcription termination in mitochondria, thus producing general defects in mitochondrial protein synthesis. In another family, a 10.4 kb deletion of mtDNA appears to be responsible for diabetes and sensory-neural deafness. Although it can be argued that the production of insulin by the beta cells of pancreatic islets is a process requiring efficient oxidative phosphorylation, that hypothesis does not fully account for the specificity of these defects. It is quite possible that cell-specific factors, perhaps encoded by polymorphic genes in the nucleus, are involved in the causation of clinical diabetes.

Mitochondrial Diseases and Nuclear DNA Mutations.

The identification of nuclear genes whose mutations disrupt mitochondrial functions is in its early stages, although rapid progress is being made. Because nearly all mitochondrial proteins are encoded by nuclear genes, it is predictable that most mitochondrial disorders will be attributable to nuclear genes. Inheritance of those disorders should follow Mendelian patterns. However, the relative frequency of mitochondrial diseases caused by mutations in mtDNA is expected to be high, because of the high mutation rate of mtDNA.

Because *oxidative phosphorylation (ox-phos) is the central function of mitochondria*, we will briefly survey disorders of that process that have been shown to be caused by nuclear gene mutations. Estimates of the prevalence of disorders of ox-phos have been in the range of 1/10,000 or slightly higher. Many of those disorders are termed *encephalomyopathies*, because they involve the nervous and muscular systems, although the endocrine system can also be affected. In general, it has been observed that the more a tissue depends upon ox-phos for its normal functions, the more sensitive it will be to disruptions of energy metabolism. The major components of the respiratory chain and ATP synthase, which together carry out ox-phos, were diagrammed in Figure 5-3.

Complex I (NADH-Q-reductase) is the first and largest of the respiratory chain complexes, composed of 43 subunits, seven of which are encoded by mtDNA. The roles of most subunits are poorly understood. Deficiencies in complex I are the commonest respiratory chain defects. Many patients have *Leigh Syndrome (LS)*, which is an early-onset neurodegenerative condition characterized by psychomotor retardation and brainstem or basal ganglia dysfunction. It is usually fatal. At least five nuclear genes that encode subunits of complex I have been shown to be defective in patients with LS. Other information strongly indicates that many cases of LS may be caused by defects in genes that encode assembly factors or maintenance factors for the complex, but do not form part of the structure of complex I.

Complex II (succinate dehydrogenase, not shown in Figure 5-3) is not part of the respiratory chain proper, but it can contribute electrons to complex III. It is composed entirely of subunits encoded by nuclear genes. Some cases of autosomal recessive Leigh syndrome have been shown to be caused my mutations in complex II genes. Mutations in two complex II proteins cause an autosomal dominant disorder called *hereditary paraganglioma*, characterized by the presence of benign tumors of parasympathetic ganglia in the head and neck. Some other cancers arise from other mutations in complex II genes.

Complex III (Q-cytochrome c reductase, containing cytochromes b and c1) catalyzes the transfer of electrons from ubiquinone (coenzyme Q) to cytochrome c. All but one of its 11 subunits are encoded by nuclear genes, but as yet no mutations in nuclear genes have been shown to be responsible for ox-phos defects. Mutations in the mtDNA gene, cytochrome b, have been associated with myopathy.

A good example of the genetic complexities that can arise from the existence of respiratory chain complexes composed of both mitochondria-encoded polypeptides and nucleus-encoded polypeptides is provided by *cytochrome oxidase* (*complex IV*, containing cytochromes a and a3). This protein is the terminal complex of the respiratory chain; its function is to transfer electrons from ferrocytochrome c to molecular oxygen. There are three mitochondria-encoded subunits, which form a core complex with catalytic activity. In bacteria, cytochrome oxidase (CO) contains only those three polypeptides, but in mammals there are also 10 nucleus-encoded subunits, whose function is still somewhat obscure. But that's not all! Mutational analysis of yeast has revealed at least two dozen other nuclear genes that are required for the assembly and full function of CO. The products of those genes affect all stages of CO assembly, beginning with processing the mitochondria-encoded mRNAs for the three core subunits, translation of the core subunits, insertion of those hydrophobic proteins into the inner membrane, addition of the heme and metal prosthetic groups, transport of nucleus-encoded proteins from the cytosol to the inner mitochondrial membrane, and all other details of the complicated assembly process of mature CO.

With so many genes contributing either to the structure or to the assembly of cytochrome oxidase, it is not surprising that CO deficiency is a frequent cause of respiratory chain defects in humans. Clinical features are quite heterogeneous, which is probably attributable to several factors. First, many mutations only suppress CO activity partially; second, there appear to be tissue-specific differences in the abundance of some CO-related proteins; and third, heteroplasmy has unpredictable effects on the abundance of mutated mitochondrial genes in different parts of the body.

Although mutations in the three core subunits (mitochondria-encoded) have been reported in humans, so far there are no cases of mutations in the other 10 subunits of the mature complex. This is puzzling, because studies on yeast have shown that all of the nucleus-encoded polypeptides of CO are required for either assembly or stability of the complete complex; they are not dispensable. However, mutations in some of the other genes involved in the assembly of CO, but not in its final structure, have been shown to be responsible for cytochome oxidase deficiency in several patients. They include mutations in proteins required for the addition of farnesyl groups, which is part of the maturation of heme A, the unique heme in the CO complex; mutations in genes required for the addition of the two copper atoms that are also part of CO; and mutations in a gene called SURF1, which appears to be required for either assembly or maintenance of CO. Evidently, there's a lot left to be learned about the effect of nucleus-encoded genes on cytochrome oxidase and the health of patients who have CO deficiencies.

Complex V, the ATP synthase, consists of 16 subunits, of which only 2 are encoded by mtDNA. One of the subunits synthesized in mitochondria (ATP6) is mutated in a syndrome known as NARP (neuropathy, ataxia, and retinitis pigmentosa) and also in some patients with Leigh syndrome. Although no mitochondrial defects have yet been associated with mutations in nuclear genes, they will inevitably be found.

There are also Mendelian disorders that have indirect effects on oxidative phosphorylation. A prominent example is *Friedreich's ataxia*, a progressive neurodegenerative disorder with a frequency of about 1 in 30,000. Expansion of a triplet repeat within an intron in the *frataxin* gene inhibits transcription. Deficiency of frataxin leads to reductions in the activity of mitochondrial enzymes with Fe-S centers, especially complexes I and III. Another example of indirect effects includes several *nuclear-mitochondrial communication disorders*, which result in loss of integrity of the mitochondrial genome. An example is *progressive external ophthalmoplegia (PEO)*, which is usually an autosomal dominant trait, characterized by exercise intolerance with onset in the third and fourth decades of human life. At least four nuclear genes have been located by linkage analysis; all of them seem to be involved with mitochondrial DNA synthesis in various ways, such that aberrant forms of their protein products lead to the accumulation of multiple mtDNA deletions.

Mitochondria and Aging

No one likes to think of aging as a disease, but we cannot escape the fact that aging is characterized by a progressive decline in many metabolic functions. Energy metabolism is a prime candidate for a major causative role in aging. For the past dozen years there has been active speculation about the role of mtDNA mutations in aging. There is no doubt that mutations in mtDNA accumulate as a person ages, and there is considerable logical appeal to the hypothesis that reduction of oxidative phosphorylation will lead to subnormal cellular functions and/or to cellular death, which will ultimately lead to organismal death. However, do mtDNA mutations *cause* any of the physiological aspects of aging, or are the mtDNA mutations a *result* of some more basic aspect of cellular function in elderly people? It's not easy to answer that question, nor is there any clear definition of *how much decline* must occur in oxidative phosphorylation before a cell or tissue becomes unable to function normally. The high level of *polymorphism* in mtDNA—a consequence of the high mutation rate—will also make it difficult to establish a causal relationship between a given mitochondrial DNA variant and a specific aspect of aging. It will take some time to sort out these uncertainties.

Therapy of Mitochondrial Diseases

At the present time, almost nothing can be done for patients suffering from diseases caused by mutations in mtDNA. Pharmacological treatments with various agents (such as ubiquinone or ascorbic acid) are at best palliative. Is gene therapy possible? The difficulties are formidable. First, getting access to the most severely affected tissues *in vivo* will be just as difficult as it is for gene therapy of diseases caused by mutations in nuclear genes. In addition, getting some sort of therapeutic macromolecule into mitochondria presents special problems, because there are special transport mechanisms for passage through the mitochondrial membranes. One suggestion has been to introduce a gene into the nucleus that codes for a protein ordinarily encoded by mtDNA. The gene could be modified so that it also encodes the N-terminal sequence of amino acids required for binding to mitochondria and being imported into them. However, we don't know whether those extremely hydrophobic polypeptides would behave docilely, fulfilling our expectations. In addition, many mtDNA mutations affect tRNAs, which in turn affect the synthesis of more than one mitochondrial protein, all of which would have to be replaced by cytosolic synthesis. Other schemes that have been considered involve DNA molecules attached to membrane-transit sequences of amino acids, which might enable the DNAs to get into the mitochondrial matrix and be transcribed there; or designing oligonucleotides that could enter the mitochondria and inhibit the replication of mutant mtDNAs, while permitting nonmutant mtDNA to replicate normally. None of these ingenious schemes shows much promise of becoming clinically useful in the near future.

REVIEW

The 16.6 kb of DNA in the human mitochondrial genome encodes 13 polypeptides and two dozen RNAs, all of which are essential components of the cellular energy production system. Those three dozen mitochondrial genes are the remnants of a complete bacterial genome, most of which has either been lost or transferred to the eukaryote nucleus, in the billion years since a bacterium became a permanent resident in the cytoplasm of a primitive eukaryote. Mitochondria still contain hundreds of proteins, but nearly all of them are encoded by nuclear genes. The proteins that carry out DNA and RNA synthesis, protein synthesis, fatty acid oxidation, the citric acid cycle, and the numerous protein constituents of mitochondrial membranes are all encoded by nuclear genes, as are various portions of the apparatus for oxidative phosphorylation.

Defects in mitochondrial genes are responsible for a variety of human diseases, all of which reduce or abolish the production of ATP via the electron transport chain from the products of the citric acid cycle. Because sperm mitochondria almost never enter a fertilized ovum, inheritance of mitchondrial genetic diseases is maternal; affected females can transmit their condition to all progeny, whereas affected males almost never transmit. Mitochondrial inheritance is complicated by heteroplasmy, which refers to the fact that that the ratio of mutant mtDNA to normal mtDNA can vary among the offspring of one affected female and can also vary from one tissue to another or from one developmental stage to another within an individual. Heteroplasmy arises because mitochondria are not precisely distributed to daughter cells at cytokinesis, in contrast to the equal distribution of nuclear chromosomes at mitosis. Heteroplasmy can also develop from differential replication of mutant mtDNA during the life span of an individual.

Because at least 95% of the proteins in mitochondria are encoded by nuclear genes, there are also many diseases affecting mitochondrial functions that show Mendelian inheritance.

Mitochondrial DNA acquires mutations about ten times as fast as nuclear DNA for several reasons, including lower fidelity of DNA replication and less accurate repair mechanisms. However, the main source of mutations in mtDNA is probably reactive oxygen species, byproducts of energy transduction that are generated within mitochondria, which readily modify DNA. It is widely suspected that the inevitable accumulation of mutations in mtDNA is a significant component of the aging process.

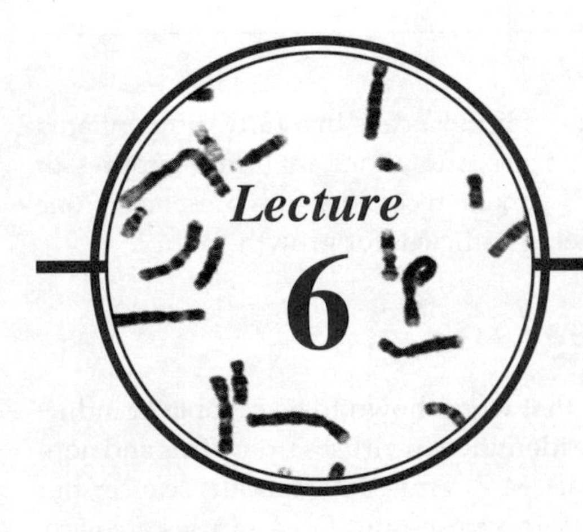

The Genetic Basis of Cancer

Cancer is fundamentally a genetic disease. Most cancers result from mutations in DNA, although epigenetic changes can also contribute to unregulated growth. Most cancers appear to be the result of somatic cell mutations exclusively, but some cancers are caused, in part, by inherited mutations; that is, by mutations transmitted through the germ cells from one human generation to the next. A few human cancers are the result of viral infection; for example, hepatitis B virus can cause hepatic carcinoma; papillomavirus (HPV) can cause cervical cancer; Epstein-Barr virus can cause Burkitt's lymphoma; and human T-cell lymphotropic virus I (HTLV-I) can cause adult T-cell lymphomas in a small fraction of infected persons. Nevertheless, the large majority of human cancers arise from endogenous mutations, not in response to any type of infectious agent.

The words "neoplasm," "tumor," and "cancer" are routinely used interchangeably to refer to any abnormal accumulation of tissue with no obvious beneficial function, but "neoplasm" and "tumor" (which are essentially synonymous) are really broader terms than "cancer." A tumor can be benign if it does nothing more than create a clump of cells that does not interfere with the function of adjacent normal tissue. To become a cancer in the clinical sense, to become *malignant*, a tumor must acquire *invasiveness*—the ability to spread into surrounding tissue and interfere with normal function. Most life-threatening tumors eventually develop the ability to form colonies (*metastases*) in many sites throughout the host's body. Nonmalignant tumors can also be life threatening if they interfere with normal functions, as is the case for some brain neoplasms, for example.

The molecular mechanisms that determine whether a cell will grow and divide are complex, and they can be disrupted in various ways. In some cases, a mutation in one gene is sufficient to lead to unregulated growth; in other cases, two mutations may be required. However, studies have shown that cancer formation in humans increases with age at a rate that implies four to six mutations are required to produce an abnormal growth that is capable of growing indefinitely and spreading to new locations within its host—a malignant cancer. If mutations occurred in tumor cells at the same rate they occur in normal cells, it would require several human lifetimes for a single cell to incur half a dozen mutations in genes that affect growth contol. Obviously, cancer cells must acquire mutations at a considerably greater rate than normal cells. Indeed, *genetic instability has been a recognized aspect of cancer cells for many years*. You can look at cancer cell chromosomes in the microscope and see many abnormalities, and molecular analyses demonstrate lots of mutations that cannot be detected microscopically. Current research is making progress toward explaining the connection between loss of cellular growth control and overall genetic instability, but it still is only vaguely understood in many cases.

Tumor development is often described in terms of stages known as *initiation*, *promotion*, and *progression*. Initiation refers to the first mutagenic event or events that give a cell the potential for unrestrained growth and multiplication. Depending upon the type of cell in which it occurs, the initiating mutation may be expressed immediately (as might be the case in a continuously proliferating stem cell population) or after a latent period of many months or years. In the latter case (which might apply to a quiescent cell in a tissue where there is very little cell division), expression of the initial oncogenic mutation may require "promotion," that is, receipt of a *mitogenic signal*; the latter might come from natural causes, such as the death of nearby cells, or from artificial causes, such as exposure to an organic chemical that has growth-promoting properties.

Acquisition of *invasiveness*, which allows a tumor to extend into the surrounding normal tissue and ultimately to interfere with normal function in that area, requires additional mutations. Tumor growth to a size much greater than a few millimeters, invasive or not, also depends upon *angiogenesis*, the development of a blood supply within the tumor. The final stage in tumor progression is *metastasis*, which occurs when cells from the original tumor migrate to new sites and establish secondary tumors there. The new sites may be in tissues quite distinct from the one in which the primary tumor arose.

Tumor progression is a multistep process requiring several somatic mutations before a full-blown metastatic cancer arises; it is not yet fully understood at the molecular level. The necessary mutations may be different in tumors that arise in different cell types. Moreover, different metastases from the same primary tumor may not have exactly the same genetic alterations. You have probably heard that most cancers are *clonal* in origin—that a given cancer develops from a single abnormal cell. That's true, if you only consider the initiating mutation. However, as genetic instability develops in a clonal primary tumor, different secondary mutations can occur in different tumor cells. Most of the primary tumor may never become invasive or metastatic, but those tumor cells that do acquire the ability to invade neighboring tissue and/or to establish colonies elsewhere are not necessarily clonal in regard to the later mutations. This can be a problem for therapy.

In the sections that follow, we will survey what is known about the genetic changes that give cancers their distinguishing features. There is more known about tumor initiation than about tumor progression, so the main topic will be the mutations that permit a cell to escape growth controls. One of the most significant generalizations that has emerged in the past 20 years is that there are two major classes of cancer-initiating genes. First, there are *oncogenes*, or *growth-promoting genes*, whose normal activity is necessary for cells to grow and divide. Mutations in oncogenes tend to be phenotypically dominant; that is, a mutant oncogene will stimulate continuous growth of a cell, despite the presence of a normal allele, which only stimulates growth at appropriate times. The second class of cancer-causing genes is a group called *tumor-suppressor genes*, sometimes referred to as *anti-oncogenes*. These are genes whose normal function is to prevent cells from multiplying, as is usually necessary for most cells in a fully differentiated organ. In most cases, mutations in tumor-suppressor genes behave as genetic recessives, the presence of one normal allele being sufficient for growth control.

ONCOGENES

The first genes that were shown to be capable of inducing cancer were identified in viruses from birds and nonhuman mammals. Most viruses do not cause cancer, but some types of RNA viruses and DNA viruses are *oncogenic*. One class of RNA viruses—the *retroviruses*—has been especially significant in the search for the genetic basis of cancer. We discussed retroviruses in Lecture 4 in connection with gene therapy. You may want to review that section. For now, the main point is that the RNA genome of retroviruses is copied into DNA inside the host cell and inserted into the host's genome, where it remains and may be expressed indefinitely.

Based on their oncogenic potential, retroviruses can be classified in two groups. *Acutely transforming viruses* induce tumors in animals within a few weeks and transform cells in culture efficiently. *Weakly oncogenic* viruses induce tumors in animals only after long latent periods (months or years) and rarely or never transform cells in culture. *Rous sarcoma virus (RSV)* is the prototype acutely transforming retrovirus and *avian leukosis virus (ALV)* is the prototype weakly oncogenic virus.

Oncogenes were discovered as a result of efforts to determine the genetic basis of the difference in oncogenic activity of RSV and ALV. An early observation was that the genome size of RSV is about 10 kb, whereas that of ALV is only about 8.5 kb. This suggested that RSV contains genetic information that is not present in ALV—a gene not needed for viral replication, but required for transformation of host cells. Because the tumors induced by RSV are sarcomas, the name *src* (pronounced "sark") was given to that gene. Further analysis showed that *src* lies between the end of the *env* gene and the 3 prime LTR (Figure 6-1).

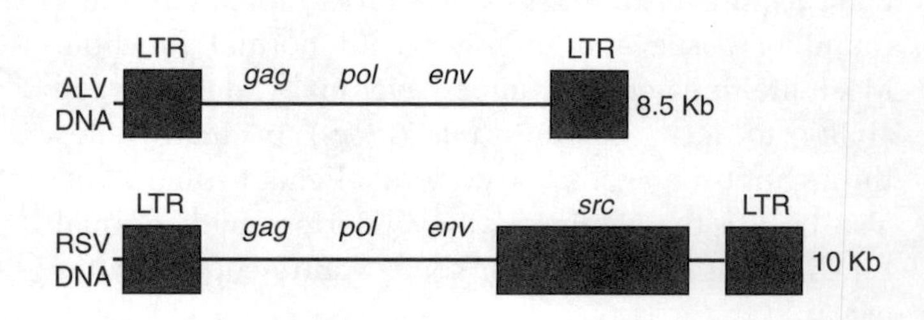

Figure 6-1 The genetic structure of two retroviruses: ALV and RSV. LTR = long terminal repeat.

RSV is an unusual retrovirus because it carries a complete set of normal retroviral genes and a separate oncogene; RSV is *replication competent*. Most other oncogenic retroviruses are *replication defective*; that is, they require the presence of a helper virus to provide one or more missing functions. In the most common situation, part of the retroviral genome has been deleted, and a gene that is necessary for transforming ability has been joined to the remaining fragment of a retroviral gene. The strong promoter associated with the retroviral genome supports active transcription of the oncogene. The product of such a gene is a *fusion protein* (Figure 6-2).

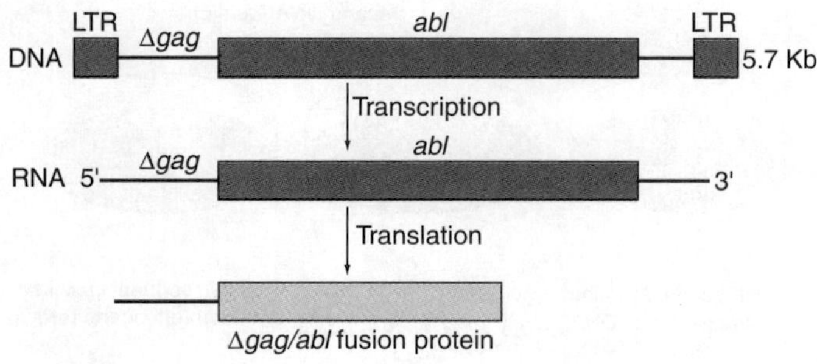

Figure 6-2 The *abl* oncogene and its product.

The fact that retroviral oncogenes are not required for viral replication raises the question of their origin and the manner in which they became incorporated into retroviruses. Evidently, the host cell genome is the most likely source of nonviral genetic information. The fact that a viral oncogene was related to a gene in normal cells was first established by Varmus and Bishop (both received the Nobel prize) with the *src* gene of RSV. Taking advantage of the existence of deletion mutants, they created a *src*-specific probe. When that probe was tested with DNA from normal chickens, it hybridized quite well. This showed that normal DNA contains sequences closely related to a retroviral oncogene.

Subsequently, all retroviral oncogenes that have been tested have been shown to have homologs in normal host cell DNA. Those normal genes have been given the name *proto-oncogenes* to indicate that they may become oncogenes when mutated, even though their normal functions are beneficial to growth control. A plausible scenario that explains the origin of retroviral oncogenes is diagrammed in Figure 6-3, which shows a retrovirus that happens to have integrated into the host genome adjacent to a proto-oncogene. Either because of a DNA deletion or because of aberrant transcription and RNA splicing, a fusion transcript is produced. If the fusion transcript is packaged into a virion together with a normal retroviral transcript (retrovirus virions typically contain two genomic RNAs), recombination may occur after the virus infects an appropriate host, creating a retrovirus that can integrate the new genetic combination into a host genome.

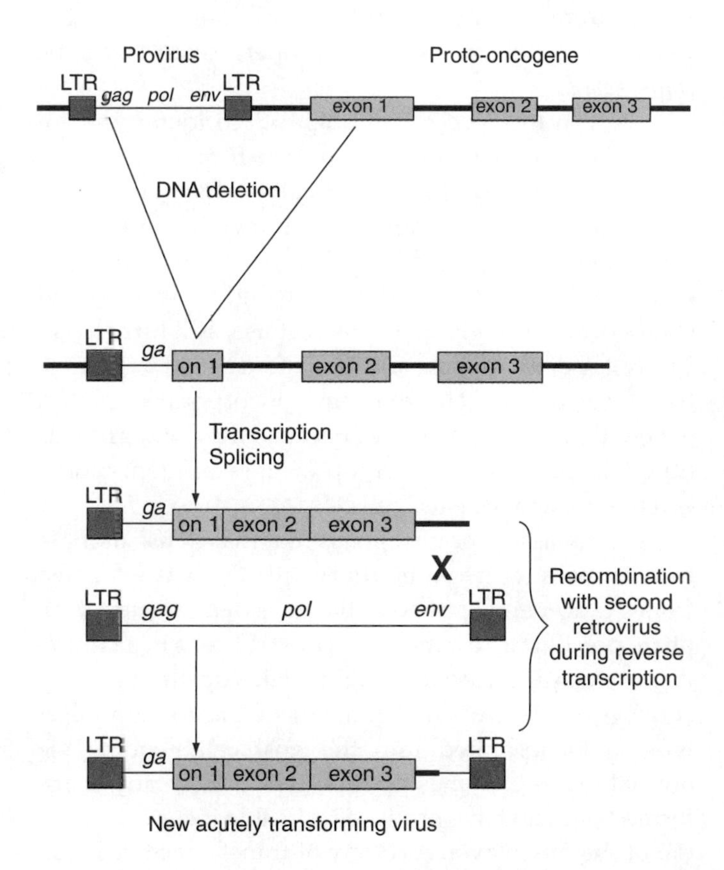

Figure 6-3 Model for generation of acutely transforming retroviruses by acquisition of an oncogene.

The existence of genes in normal cells that are similar to retroviral oncogenes raised the possibility that tumors could arise by spontaneous alterations in the structure or expression of those genes, without the mediation of viruses. It therefore became important to examine human tumors for evidence of endogenous alterations in known proto-oncogenes. Several experimental approaches were taken, and several types of alterations in proto-oncogenes were discovered that convert the proto-oncogenes into *cellular oncogenes*. In the course of that work, many proto-oncogenes have been discovered that are not related to any known viral oncogene.

One major technique relied on the plausible assumption that an oncogene, if it is present in DNA extracted from a cancer, can be detected by transferring it to suitable cells in culture, which should then show loss of growth control. Tumor cells in culture share two invariant properties with tumor cells in an organism: *immortality* (that is, no limit on the number of cell divisions that can occur if space and nutrients are available) and *uncontrolled growth* (which, in culture, is expressed as an escape from density-dependent inhibition of growth). Cells that exhibit both of those properties are said to be *transformed*.

The first successful attempts to identify oncogenes from human tumors via transformation of cultured cells employed NIH 3T3 cells, which are mouse cells that have been immortalized, but which still exhibit *contact inhibition* (i.e., they will form a *monolayer* in a culture vessel and then stop growing). It was a good choice of experimental systems because, as it turned out, 3T3 cells need only one additional oncogene to become transformed and cells that have incorporated such a gene will be easily detected, because they will grow as little *foci* (piles of cells) on top of the confluent monolayer of growth-arrested cells.

An assay for oncogenes that makes use of these facts is really quite straightforward. DNA is extracted from a cancer and applied in the form of a calcium phosphate precipitate to a monolayer of 3T3 cells in a culture dish, which induces the cells to take up the DNA by endocytosis. Some small fraction of the foreign DNA will be incorporated into the host cell genome via normal repair enzymes. Several weeks later, any transformed cell will have produced a visible focus of cells on top of the monolayer. A colony of transformed cells can be isolated and multiplied to any desired level; then, DNA can be extracted and used to initiate another cycle of transformation. This is done so that the probability of there being more than one piece of tumor cell DNA integrated into the recipient cells will be very low.

The human gene that was responsible for the transformation of the mouse cells can then be isolated, after DNA from the transformed cells has been used to make a clone library (a collection of DNA fragments cloned into bacteria), by identifying the clones that contain human-specific Alu sequences (see Lecture 2). These sequences are so common that most pieces of human DNA 20 kb or larger will contain at least one.

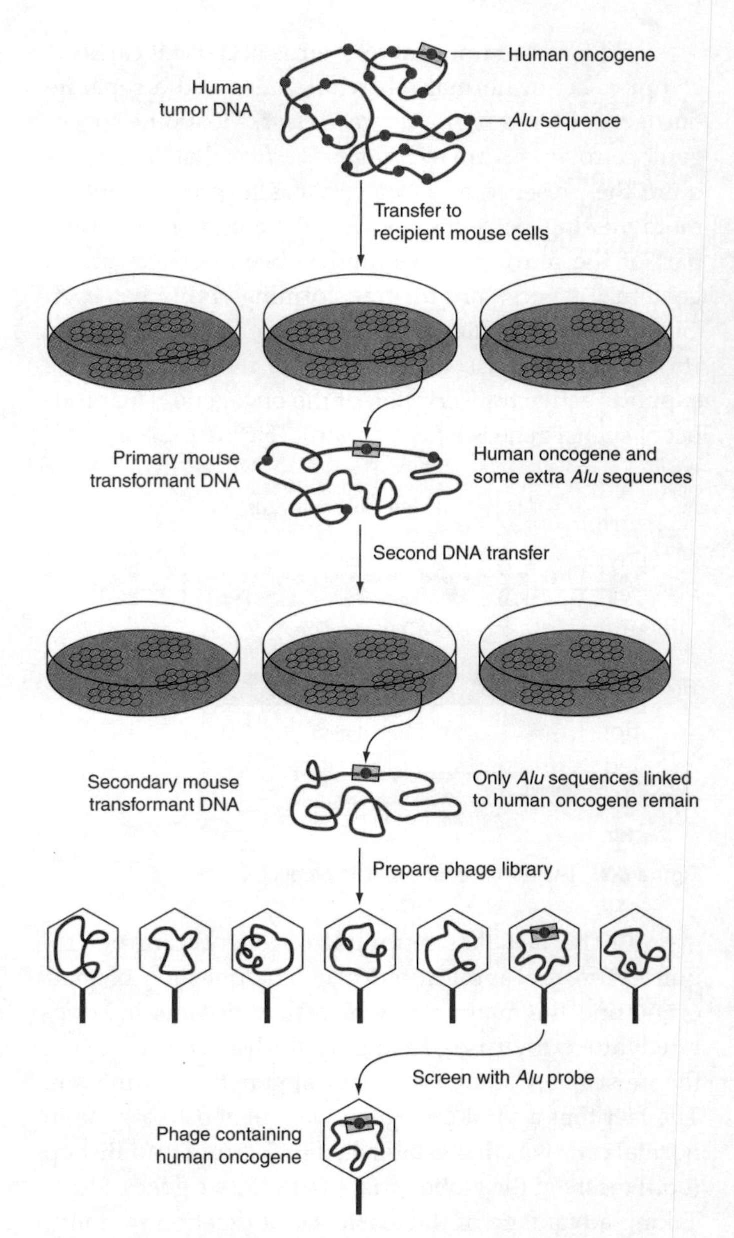

Figure 6-4 Molecular cloning of human oncogenes.

The first cellular oncogene isolated by the gene transfer strategy was derived from a human bladder carcinoma cell line. Subsequently, it became apparent that this oncogene was related to the *ras*H oncogene of Harvey sarcoma virus (a mouse retrovirus). Another cellular oncogene, initially isolated from a human lung carcinoma, proved to be homologous to *ras*K, from Kirsten sarcoma virus. A third member of the cellular *ras* gene family is *ras*N, so named because it was first isolated from a neuroblastoma.

Many such experiments have been done, using DNA from a variety of spontaneous tumors, such as carcinomas of the bladder, colon, and lung, as well as sarcomas, neuroblastomas, and leukemias. DNA from tumors produced by carcinogens in the laboratory has also been used. The general conclusion is that many tumors, regardless of origin within the body, have alterations in their DNA that will cause transformation of suitable fibroblasts in culture.

However, only about 20% of human cancers contain DNA that will transform 3T3 cells. This is believed to be a limitation of the assay, rather than evidence that no oncogene is present. The 3T3 cell transformation assay detects only mutant genes that are phenotypically dominant and are able to complement the effects of an endogenous mutation that has made the cells immortal.

Most of the oncogenes identified by the gene transfer assay are derived from proto-oncogenes by a mutation that changes the amino acid sequence of the encoded protein (a missense mutation) or changes a nucleotide involved in regulating expression of that gene. Another important source of oncogenes is *chromosomal rearrangements*. Chromosome instability is one of the most conspicuous characteristics of neoplastic cells. In most tumors, the karyotypic abnormalities are not predictable and they may be late consequences of transformation. However, some tumors appear to be the result of specific chromosomal rearrangements.

The original example involved *chronic myelogenous leukemia (CML)* and the "Philadelphia chromosome," a short version of chromosome 22 that is actually the product of a reciprocal translocation between small pieces at the ends of the long arms of chromosomes 9 and 22 (Figure 6-5). At least 90% of CML patients have the Philadelphia chromosome in their leukocytes. Molecular analysis showed that most of the proto-oncogene *ABL1* from 9q is joined to a gene called *BCR* ("breakpoint cluster region") on 22q. The fusion protein expressed by this compound gene is a *tyrosine kinase* with much greater activity than normal *ABL1* gene products. Many tyrosine kinases, which phosphorylate the amino acid tyrosine at specific sites in certain proteins, are crucial parts of signaling cascades that are involved with the regulation of cell growth and gene expression. We now have a list of at least three dozen oncogenes produced by chromosomal rearrangements.

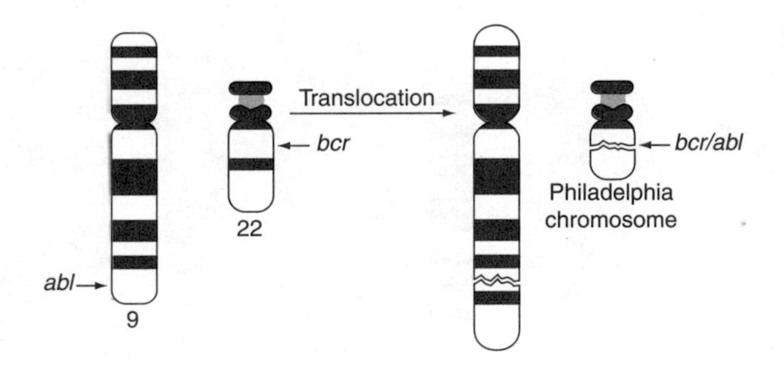

Figure 6-5 The Philadelphia chromosome.

The last mechanism for the origin of oncogenes that we will consider is *gene amplification*. It is often observed that a tumor gradually becomes insensitive to a chemotherapeutic agent that was initially effective. Molecular analysis has shown that drug resistance is frequently the result of gene amplification. The classic example involves resistance to methotrexate, an inhibitor of dihydrofolate reductase (an enzyme essential for DNA nucleotide synthesis). Tumors that become resistant to methotrexate can usually be shown to have multiple copies of the dihydrofolate reductase gene—sometimes as many as several hundred copies per cell.

Amplified genes are usually organized in tandem arrays, which may be detected intrachromosomally as *homogeneous staining regions* or as *double minute chromosomes* (Figure 6-6). The latter are independent minichromosomes that lack centromeres. They are distributed randomly at mitosis, and they are maintained in a population of cells only if they confer a growth advantage to those cells that possess them.

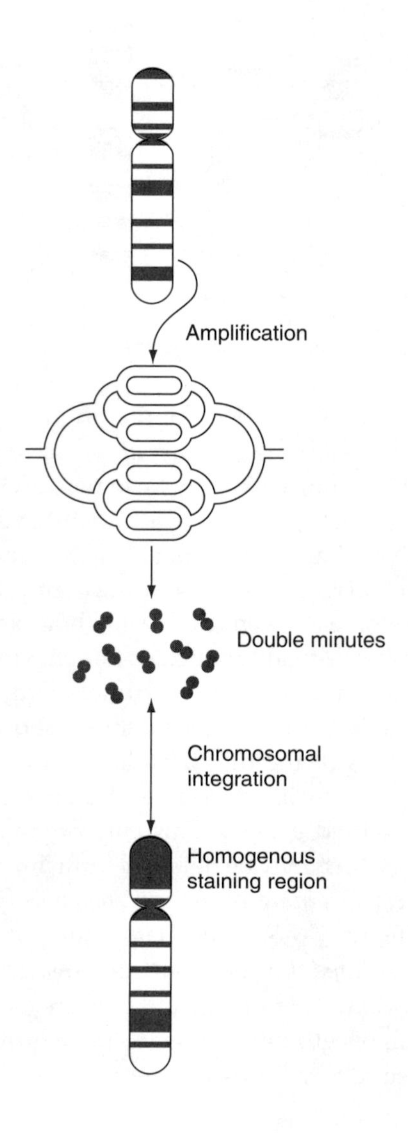

Figure 6-6 The origin of amplified genes.

More than a dozen oncogenes have been found to be amplified in human tumors. In most cases of amplified genes, it is the proto-oncogene that is amplified, and the normal protein is produced. (This leads to a semantic problem, because we have been defining an oncogene as a mutant form of a proto-oncogene, but we'll just have to accept this little inconsistency.) Presumably, an excessive amount of certain proteins deregulates growth control mechanisms, resulting in unrestrained cell growth. In some tumors, an amplified oncogene is found that has also undergone a mutation that may enhance its transforming potential. There is considerable uncertainty as to whether gene amplification is a primary or secondary event in tumorigenesis.

TUMOR-SUPPRESSOR GENES

In the preceding section, we discussed oncogenes—mutant forms of normal genes (proto-oncogenes) that act in a dominant negative manner. One copy of an oncogene is often sufficient to initiate the process of unregulated cell growth in an otherwise normal cell. Alternatively, when introduced into a suitable recipient cell, some oncogenes cause that cell to become transformed, to exhibit continuous density-independent growth in culture, despite the presence of two copies of the corresponding proto-oncogene. We shall now consider a different class of oncogenes—those whose normal function is to inhibit cell division unless and until the rest of the regulatory system signals that cell division is appropriate. When both copies of one of these negative regulators are lost or inactivated in a given cell, unrestrained growth ensues and a tumor may develop. These genes have been called anti-oncogenes and recessive oncogenes, but *tumor-suppressor genes* is by far the most common name for them.

The first evidence for the existence of genes that could suppress unrestrained growth of cancer cells came from studies on hybrids between a cancer cell and a normal cell in a mouse system (somatic cell hybrids were mentioned in Lecture 1). Most of the combinations had normal growth control in culture, and when transferred to mice, no tumor developed. Moreover, in the case of certain hybrid cells, the loss of specific chromosomes correlated with reversion to the transformed phenotype. Additional studies in which single chromosomes from normal cells were transferred to tumor cells in culture showed that suppression of tumorigenicity could be correlated with specific chromosomes. Those observations implied that one way in which a tumor cell might arise would be the loss of normal function of a growth suppressor gene. Molecular studies have now confirmed that hypothesis in many systems and at least two dozen tumor-suppressor genes have been identified.

The best-known example of a tumor suppressor gene is the *retinoblastoma gene, RB1,* which has an important role in control of the cell cycle (as discussed later) and a very interesting history. Retinoblastoma has long been known as a rare form of inheritable cancer, found in about 1 child in 20,000. Tumors in one or both eyes develop in the retina within the first few years of life, and if untreated, spread along the optic tract to the brain and become lethal. Currently, a tumor detected in its early stages can be removed or destroyed with a laser, with complete cure ensuing. There are both *inherited* and *sporadic* (noninherited) forms of retinoblastoma, with the latter representing somewhat more than half of the total cases. In inherited retinoblastoma, the disease is passed from generation to generation as an apparent autosomal dominant, with about 90% penetrance.

Children with inherited retinoblastoma usually develop more than one tumor in each eye, and the tumors appear earlier than in sporadic cases, where there is generally only one tumor in one eye. These and other data on the pattern of inheritance led Knudson in 1971 to suggest that two mutations are necessary for the appearance of a retinoblastoma. At that time, it was not possible to predict whether the two mutations would be in different genes or in the two alleles of one gene. We now know the latter to be true.

According to the *Knudson "two-hit" hypothesis,* inherited retinoblastoma occurs in individuals who have received one copy of a mutant *RB1* gene from a parental germ cell. A second mutation occurs in one or more retinoblasts during early childhood, as the retina develops its definitive structure. Individuals with sporadic retinoblastoma must have experienced two separate somatic cell mutations in a retinoblast or its precursors (Figure 6-7). This general concept applies to all cancers where the initiating event involves loss of function of a tumor-suppressor gene.

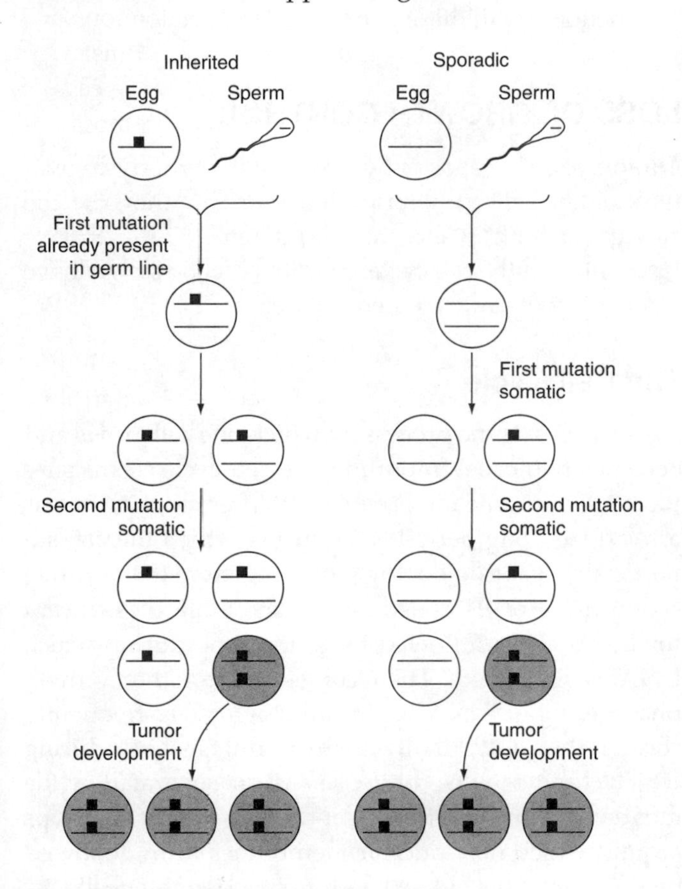

Figure 6-7 The two-hit model of tumor initiation.

The first clue to the location of the *RB1* gene came in 1978, when Yunis and Ramsay noticed that cells from a small percentage of retinoblastomas had a cytologically visible deletion in band 14 of the long arm of chromosome 13. Subsequently, it was observed that, in some patients, the deletion was homozygous in the tumor (or only the abnormal chromosome 13 was present in the tumor), although it was heterozygous in the patient's normal tissues. Comparable observations were then made on esterase D, a protein encoded by a gene closely linked to *Rb*, which also mapped to 13q14.

This information led to the hypothesis that both copies of the *RB1* gene must be missing or inactivated in order for a tumor to form. A child born with one mutant *RB1* allele, such as the interstitial deletion at 13q14, need only lose the wild type allele from one retinoblast during development of the retina, in order for a tumor to form (Figure 6-7). Several potential mechanisms for loss of the normal allele from a cell are illustrated below. Although they are rare events, the fact that there are several million retinal cells in each eye suggests that at least one such loss would be very likely, thus accounting for the high penetrance of the disease. The process of losing one allele from a cell that is already heterozygous at a given locus is called *loss of heterozygosity* (LOH). It is an important general mechanism for the development of cancers in persons who inherit a predisposing allele.

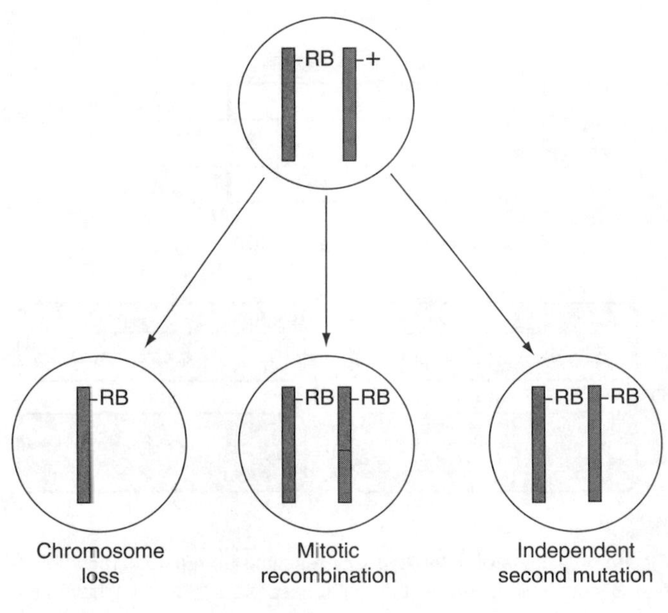

Figure 6-8 Mechanisms for loss of heterozygosity.

One well-known mechanism for loss of an entire chromosome is *nondisjunction* (see Lecture 2), which in this case would be loss of the chromosome 13 that contains the normal allele at anaphase, resulting in one daughter cell that is *trisomic* for chromosome 13 and one that is *monosomic*, with the mutated allele. In the latter case, there would be no remaining functional *RB1* allele in that cell, and it could initiate a tumor, if the monosomic cell was viable. Two other mechanisms that could lead to a cell with two nonfunctional *RB1* alles are mitotic recombination (Figure 6-9) and an independent second mutation. Both are rare events, but only one such event is needed to initiate a tumor in a person who already has one nonfunctional allele at a tumor-suppressor locus.

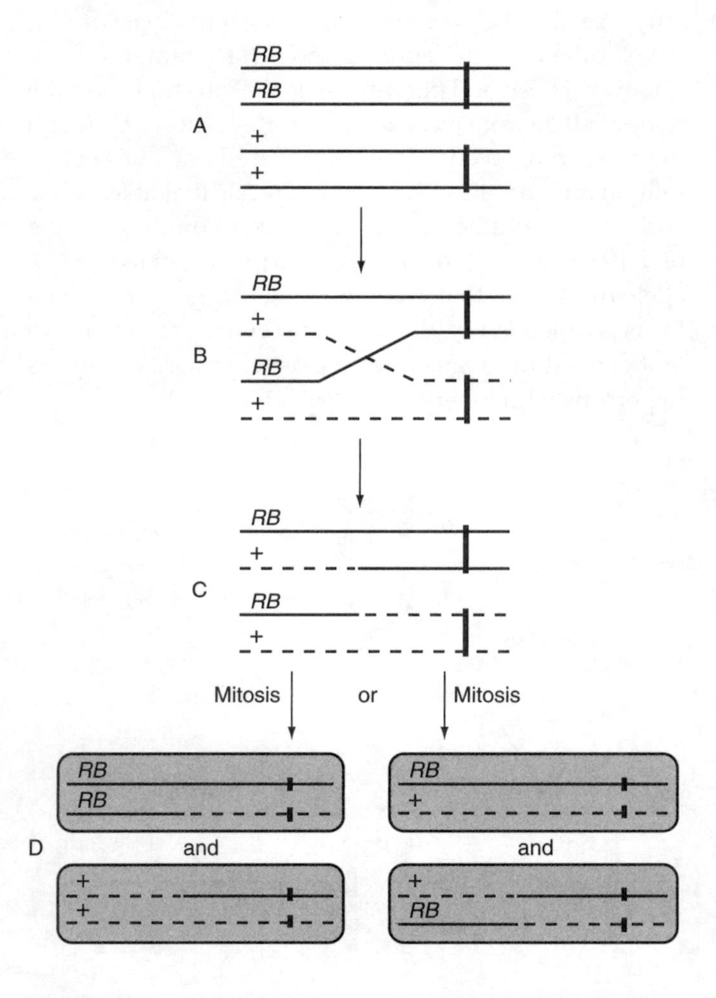

Figure 6-9 Loss of heterozygosity resulting from mitotic recombination. Because alignment on the metaphase plate is random, half of the daughter cells will no longer be heterozygous at the tumor-suppressor locus.

In contrast, it is presumed that a child with sporadic retinoblastoma must have acquired two somatic mutations, one in each *RB1* allele. Whether the first mutation increases the probability of a second mutation at the Rb locus has not been determined.

Some other examples of tumor-suppressor genes are the following:

1. *APC* is a gene expressed in intestinal epithelium; *persons with a germline mutation in one APC gene develop familial adenomatous polyposis*, which is characterized by hundreds or thousands of benign polyps developing from glandular structures (adenomas) in the colon and rectum. Acquisition of a second mutation that inactivates the remaining normal allele leads to colon carcinoma, with high probability.
2. *The gene TP53 is mutated in approximately 50% of all cancers*. Persons with Li-Fraumeni syndrome inherit one inactive allele at the TP53 locus; they have a high probability of developing one or more tumors (usually sarcomas) before age 45.
3. *BRCA1 and BRCA2 are genes frequently involved in breast and ovarian cancer*. Persons who inherit an inactive allele have up to a 90% probability of developing one of those tumors by age 80. We will consider the functions of all these genes in a later section.

LOSS OF GROWTH CONTROL

Tumors arise because of two general types of dysregulation: either cells proliferate too much or they die too slowly (or both). Cells that proliferate too much have lost control of the cell cycle; cells that die too slowly have lost control of programmed cell death.

The Cell Cycle

The cell cycle is the process by which one cell grows and becomes two cells. In outline, the cell cycle seems simple, as diagrammed in Figure 6-10. After mitosis, there is a relatively long period (G1) during which the cell superficially seems to be doing nothing more than synthesizing more of its constituent macromolecules, except for DNA. G1 is followed by S, the time during which DNA is replicated. Then comes G2, another growth phase without DNA replication, which is frequently shorter than G1. Finally there is mitosis (M), during which chromosomes condense and are separated by the movements of the mitotic spindle into identical groups around which new nuclear membranes form, followed by division of the cytoplasm into two daughter cells.

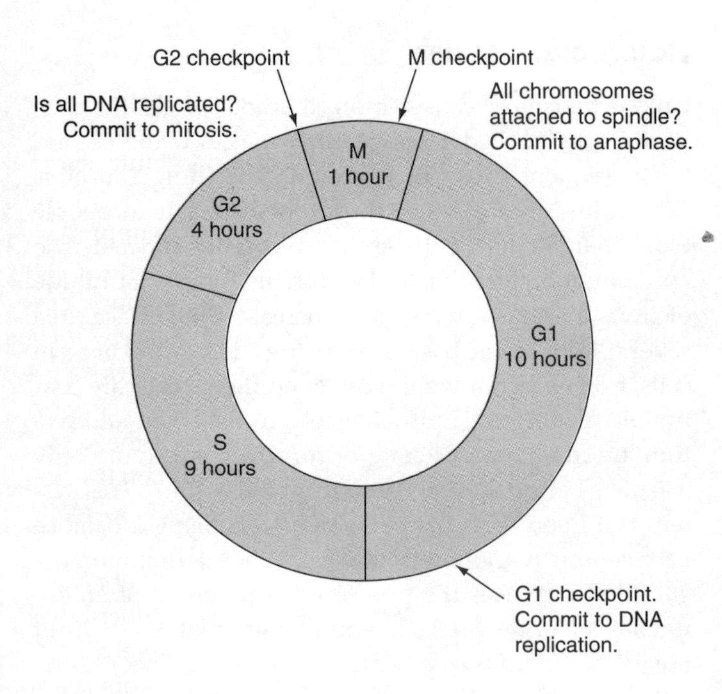

Figure 6-10 The cell cycle and its major checkpoints.

In reality, a complex system of controls exists within the cell cycle, whose most important functions are (1) to ensure that DNA is not replicated until the cell is fully prepared for that crucial genetic function, and (2) to prevent mitosis from taking place if there are any mistakes in the DNA sequence that have not yet been repaired.

A crucial event in cell cycle control is the decision to enter S phase and replicate DNA. The decision to proceed to S phase is actually made in G1, some time before the beginning of DNA synthesis, at a time called *the restriction point (R point)*. Among the proteins that regulate the R point are *cyclins, cyclin-dependent kinases (cdk's)*, and the proteins that interact with them. Cyclins are a family of about a dozen proteins, synthesized at various times during the cell cycle, that bind to cdk's and activate them. For example, D-type cyclins are required for progression through G1, and cyclin E activity peaks at the G1/S interface. Cyclins tend to have very short half-lives; their activity is a balance between transcription and proteolysis.

The full list of substrates for cyclin-dependent kinases is not yet known, but one prominent substrate is RB1, the product of *the retinoblastoma locus*. During most of G1, unphosphorylated RB1 protein binds a transcription factor called E2F, keeping it inactive. RB1 is gradually phosphorylated by cdk's, eventually releasing E2F, which is required for the transcription of a set of genes needed for S phase. E2F is subsequently inactivated by the cyclinA-cdk2 complex during S.

So you see, we already have a list of four protein classes involved in making the decision to enter S phase: cyclins, cdk's, Rb, and E2F. This is just a sample; many more genes that participate in cell cycle control are known, and there is little doubt that others will be identified. Mutations in a large number of genes can lead to improper regulation of the cell cycle and be a major factor in the initiation of a tumor.

Programmed Cell Death

Virtually all cells have the inherent capacity to commit suicide via a regulated pathway of reactions. The name *apoptosis* has been given to programmed cell death; it means "falling leaves." The primary characteristics of apoptotic cell death are blebbing of the cell membrane accompanied by cell shrinkage, dense condensation of the nucleus, and digestion of the genome into fragments that are multiples of roughly 200 bp, which is the amount of DNA associated with single nucleosomes. *Apoptosis is a response to internal signals.* It is quite distinct from *necrotic cell death*, which is a response to external trauma; necrosis is characterized by cell swelling with disruption of the cell membrane and the organelles. Necrotic cell death induces an inflammatory response; apoptosis does not.

Apoptosis plays an essential role in the elimination of surplus cells during development, such as tissue between the digits and differentiation of several parts of the brain. Apoptosis is also one of the body's prime defenses against cancer, and the loss of cell death regulatory mechanisms is one of the most common aspects of malignancy. Normal tissues maintain *homeostasis* via a balance between cell proliferation and cell death; dysregulation of either system can produce a tumor.

The central genetic elements of programmed cell death have been tightly conserved in animal evolution, and some of the most basic insights into the process came from studies on the nematode worm, *C. elegans*. We now know the identity of human genes that correspond to the apoptosis-control genes in the worm. An important example is *BCL2* (the name refers to B cell lymphoma/leukemia-2), which encodes a potent inhibitor of apoptosis. In many B cell lymphomas, *BCL2* is translocated from its normal location at chromosome 18q21 to chromosome 14, where it comes under the control of the immunoglobulin heavy gene enhancer. This produces a dramatic up-regulation of *BCL2* with the result that a clone of lymphocytes arises that do not die at the normal rate. They don't proliferate at an abnormally high rate; they just live longer than normal lymphocytes. This creates a lymphoma—a cancer of lymphoid cells. The metabolic basis is quite different from the enhancement of proliferative potential induced by oncogenes and from the inhibition of anti-proliferative signals that results from inactivation of both copies of tumor-suppressor genes.

SOURCES OF GENOMIC INSTABILITY

We noted at the beginning of this lecture that tumor progression—acquisition of invasiveness and the ability to form metastases—requires more mutations than the one or two initial changes that allow a cell to escape normal controls on the cell cycle. What causes those secondary mutations? Does escape from cell cycle regulation somehow create a *mutator* phenotype? We are a long way from a definitive answer to that question, but one widely accepted generalization is that loss of cell cycle regulation leads to various types of genomic instability. It is a fact that life at the DNA level is replete with uncertainties and dangers. There are both external and internal sources of threats to the integrity of the genome. External threats include many *DNA-modifying compounds* that may be ingested with food or inhaled and, for skin cells at least, the ultraviolet light component of sunlight can cross-link adjacent bases, creating *pyrimidine dimers*. Internal threats include *reactive oxygen species* generated as a normal part of energy metabolism, but they also include the less obvious category of mistakes during replication, repair, and recombination. Those replication mistakes are only to a minor extent related to incorporation of the wrong bases. A more serious problem is related to *stalled replication forks* (which occur for a variety of reasons) and to *illegitimate recombination* (DNA duplexes sometimes form between closely related sequences at nonallelic sites; recombination at those points leads to deletions and duplications).

A very complex set of proteins is involved with fixing all of those problems and, in one way or another, loss of cell cycle regulation usually also affects some component of the metabolic network that maintains genomic integrity. Dysregulation of programmed cell death can also feedback onto the circuits that control genomic integrity. Now let's look at a few specific examples.

Colorectal Cancer

Colorectal cancer causes more than 50,000 deaths each year in the United States, which makes it the second most frequent cause of cancer deaths (lung cancer is first). More than 90% of those deaths occur in people older than 55, and both sexes are equally affected. The epithelium of the colon and rectum includes a multitude of invaginations (*crypts*) that increase the surface area several-fold. At the base of each crypt is a group of stem cells, from which several types of epithelial cells are continually generated. Epithelial cells in the colon and rectum are in a constant state of turnover, surviving only 3–6 days before they are shed from the surface. This system is a good example of homeostasis being a balance between birth and death of cells. When a mutation occurs that increases the proliferative potential of an epithelial cell or decreases the probability that it will die at the usual time, a *neoplasm* arises. (The word "neoplasm" only refers to an abnormal accumulation of cells with a clonal origin; it does not imply a cancer.)

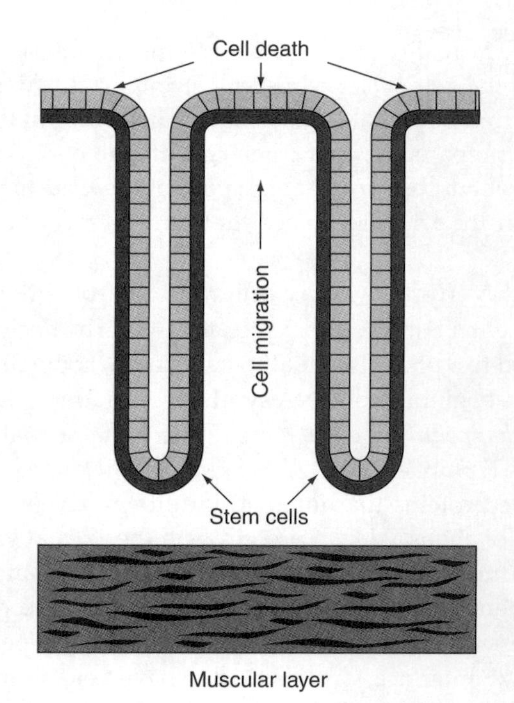

Figure 6-11 A section of colon epithelium.

Neoplastic growth in the colon first becomes evident as a polyp, a mass of cells protruding into the lumen. One type of polyp, defined by its histological structure, is called an *adenoma*; adenomas are the source of malignant colorectal cancers. Small adenomas are very common, especially in older people, but they do not necessarily imply that a cancer will develop. Adenomas larger than 10 mm are more likely to become malignant. Colorectal cancers tend to metastasize rapidly, and 40% of patients die within 5 years of diagnosis.

Several types of genetic alterations are frequent in colorectal cancer and, as is often the case, inherited predispositions to cancer in certain families have provided important clues leading to the identification of the genes involved. One syndrome is *familial adenomatous polyposis (FAP)*, which is inherited as an autosomal dominant condition. Affected persons develop hundreds (sometimes thousands) of polyps, beginning in the second decade of life and increasing for the next two decades. The median age at diagnosis for colorectal cancer is 40 years. Studies on interstitial deletions and on allelic losses in FAP patients led to chromosome 5q21 and eventually to identification of the *adenomatous polyposis coli (APC) gene*.

APC encodes a large protein (2,843 amino acids), which is located at the basolateral membrane of epithelial cells in the colon and rectum. As epithelial cells move from deep within intestinal crypts to the surface, *APC* expression increases. It is thought that this is an important element in the decision to undergo programmed cell death, which is the normal fate of those cells. This involves the APC protein's interaction with several other proteins, of which the most interesting seems to be *beta-catenin*—another multifunctional protein. Catenins are a family of proteins that bind to *cadherins*, a family of cell adhesion molecules. Beta-catenin is crucial for cell adhesion; however, beta-catenin is also part of a transcription complex that stimulates cell growth. If beta-catenin is bound to APC, it cannot bind to cadherins or to transcription factors; moreover, APC-bound beta-catenin is targeted for degradation by *proteasomes*. This rather complex system implies that *APC is a tumor-suppressor gene*, and that at least one of its functions is to control the activity of beta-catenin. When APC is mutated in such a way that it cannot bind beta-catenin, cell growth is stimulated and cell adhesion is inhibited.

The APC gene is mutated in most colorectal tumors, not just in FAP patients. The frequency of mutant alleles at the *APC* locus is just as high in benign polyps as in cancers, so loss of APC function is probably an early event, perhaps the initiating event, in colorectal tumorigenesis. This stands in contrast to the next system we shall consider, the *TP53* gene (formerly p53), which is mutated and/or lost in three-fourths of colorectal tumors, but rarely in adenomas, their noncancerous precursors. This suggests that changes in TP53 are more likely involved in tumor progression than in tumor initiation.

The *TP53* gene is a tumor-suppressor locus on chromosome 17q that is mutated in a large variety of tumors, including bladder, brain, breast, lung, and cervix. The TP53 protein forms a tetramer that binds to DNA at specific sites; it is a transcription factor that has already been shown to regulate expression of at least 20 genes, and the list is probably incomplete. The genes regulated by TP53 are involved in cell cycle regulation (especially a G1 arrest point), apoptosis, and self-regulation of the TP53 pathway. Although the functions of TP53 in normal cells are not fully understood, they probably include prevention of propagation by cells with DNA damage. This would explain why ablation of TP53 function is apparently a major factor in tumorigenesis. Mutations in TP53 occur more frequently in human cancers than mutations in any other gene.

Li-Fraumeni syndrome (LFS) is a rare (about 1 in 50,000) inherited predisposition to multiple cancers that is primarily associated with mutations in TP53. Patients develop a bone or soft tissue sarcoma before the age of 45 and typically have at least one first-degree relative who also develops early cancer. Breast cancer, brain cancer, and many other types of malignancies have also been found in families where Li-Fraumeni syndrome occurs. In view of what was said about the frequency of TP53 mutations in colorectal cancer, it is surprising that Li-Fraumeni patients do not have an unusually high frequency of colorectal cancer. This should remind us that tissue-specific factors must play an important role in determining whether a tumor will develop in response to a given mutation.

Genetically, LFS behaves as a typical Mendelian dominant condition, although biochemical evidence implies that TP53 is a tumor-suppressor gene. Most abnormal TP53 alleles have missense mutations, which may prompt you to wonder why Li-Fraumeni syndrome is not inherited as a Mendelian recessive. The answer is at least partially to be found in the fact that *TP53 functions as a homotetramer*—a protein made up of four identical subunits. Accordingly, if normal and abnormal polypeptides form tetramers on a random basis, only 1/16 of the tetramers will consist entirely of normal subunits. That level of normal TP53 function may not be adequate to maintain cell cycle regulation. Nevertheless, the TP53 story is complicated, and we can expect to learn a lot more about its roles in human cancer in the years ahead.

Breast Cancer

Breast cancer is the leading cause of death in American women in their late 40s and early 50s. Approximately one woman in eight will be diagnosed with breast cancer at some time, and nearly one-third of those women will die of the disease. Family history is a risk factor for 15–20% of women with breast cancer, but *only 5% of all breast cancers can be attributed to the best-known susceptibility loci, BRCA1 and BRCA2*. Inherited mutations in either of those genes behave like classic Mendelian dominants with very high penetrance—that is, a woman who inherits a susceptibility allele at either the *BRCA1* or *BRCA2* locus has about a 90% probability of being diagnosed with breast cancer at some point in her life, and her female children have a 50% probability of developing breast cancer eventually. Moreover, age at diagnosis is significantly lower and the probability of bilateral cancer is significantly higher than for sporadic cases of breast cancer.

The very high predisposition to breast cancer in women who inherit a mutant allele at either the *BRCA1* or *BRCA2* locus is frequently a result of *loss of heterozygosity*. Just like the situation for retinoblastoma (which was introduced earlier in this lecture), there seems to be a high probability that the remaining normal allele will be lost or inactivated in one or more mammary epithelial cells, initiating a tumor. Carriers of mutant *BRCA1* or *BRCA2* alleles also have an increased risk of developing ovarian cancer, but not as high as the breast cancer risk. *BRCA2* mutations also involve a significant risk (about 5%) for male breast cancer.

Both *BRCA1* and *BRCA2* are large genes, with *BRCA1* encoding a protein of 1,863 amino acids and *BRCA2* encoding a protein of 3,418 amino acids. More than 500 mutations have been described in each gene, which suggests that virtually the entire protein is crucial for its functions. This also implies that population-based genetic screening programs are not feasible; there are just too many possibilities to test, except in special groups where only a few cancer-associated mutations occur (as in Ashkenazi Jews). Both genes are widely expressed in the human body, so the fact that mutations in them are primarily associated with only two organs, breasts and ovaries, is still in need of an explanation.

Evidence that BRCA1 and BRCA2 proteins are involved in DNA damage repair is provided by the finding that both associate with other DNA repair proteins *in vivo*, although the precise functions of neither molecule have been determined. BRCA1 and BRCA2 also associate with each other, suggesting that they function in the same pathway. Both proteins are primarily localized in the nucleus of normal cells and both have properties consistent with a role as *coactivators or corepressors of transcription*. Both proteins associate with TP53, a protein prominently associated with many types of cancer and involved with cell cycle regulation, as explained above. Both BRCA1 and BRCA2 levels change during the cell cycle, with low levels early in G1 and the highest levels in S. All of these properties suggest complex functions for BRCA1 and BRCA2, including DNA damage repair and interactions with other proteins that regulate progression through the cell cycle. Much remains to be learned.

Cancers with DNA Maintenance Defects

So far, we have mentioned several genes whose functions are suspected to be involved with DNA repair, at least indirectly, but where the evidence is still incomplete. Now we will take a brief look at several human genetic diseases that clearly are caused by loss of function of a component of the DNA replication and repair network. Although each of those diseases has a distinct clinical presentation, all of them include increased susceptibility to cancer.

Xeroderma pigmentosum (XP) is an autosomal recessive condition characterized by exceptional photosensitivity and cataracts. Persons with XP have a 1,000-fold greater risk of skin cancer than the general population. There are at least seven genes in which mutations can lead to the XP phenotype; all of them encode proteins required for *nucleotide excision repair (NER)*—the process of recognizing a lesion in DNA, unwinding the DNA containing the lesion, cutting out the strand that contains the lesion, and synthesizing a normal piece to replace the DNA that was removed (using the other strand as a template, as usual). This process is different from *mismatch repair*, which deals only with single nucleotides.

Two of the XP genes (*XPB* and *XPD*) encode *helicases*, which are proteins that convert double-stranded DNA to single-stranded DNA by localized displacement of the hydrogen bonds that hold the strands together. This is necessary for several processes, including transcription, DNA replication, DNA repair, and recombination. Human cells contain a family of helicases, each specific for some aspect of DNA or RNA metabolism. The other XP gene products function in other steps of the NER process, including identifying damaged DNA and cleaving the damaged DNA on both sides of the lesion.

XP gene products act in concert with other proteins, including transcription factors and single-stranded DNA-binding proteins. It is estimated that at least 30 genes are involved in NER. When a person is homozygous for inactive alleles at any of the seven XP loci, nucleotide excision repair is faulty, so damage to DNA accumulates, especially in skin cells, where pyrimidine dimers are frequently generated in response to the ultraviolet component of sunlight.

Bloom syndrome (BS) is a rare autosomal recessive disorder characterized by a greatly increased predisposition to develop cancer of all types. Persons with BS tend to be short, with infertility in males and subfertility in females. They develop sun-sensitive facial redness and are very prone to become diabetic. At the cellular level BS shows a high level of mutations at all loci, plus chromosomal rearrangements and a uniquely high frequency of *sister chromatid exchanges*. The gene (called *BLM*) encodes a large protein that belongs to a family known as *RecQ helicases*. It is expressed in all tissues, but at higher levels in thymus and testis; cell cycle studies show that expression of *BLM* peaks during S phase. A molecular explanation for the BS phenotype is not yet available, but studies on model organisms strongly suggest that all the RecQ helicases are involved in resolution of structural abnormalities that arise during DNA replication, such as stalled replication forks, and in correction of illegitimate recombination.

Werner syndrome (WS) is another rare autosomal recessive disorder, well-known as a premature aging syndrome. Affected persons develop normally during childhood, but after puberty, they begin to show early onset of atherosclerosis, osteoporosis, hair loss, and cataracts. They also have a marked predisposition to diabetes and certain forms of cancer. Cells from WS patients have a reduced lifespan in culture and show a peculiar form of genetic instability characterized by a wide variety of translocations. The gene that underlies WS, called *WRN*, encodes another helicase of the RecQ family. However, the WRN and BLM proteins differ from each other in nonhelicase portions of the molecule; in particular, WRN also possesses an exonuclease activity that is not present in BLM. We do not yet know whether the differences in BS and WS phenotypes result from different sites of action of the helicases, from different activities of the nonhelicase portions of those large proteins, or both.

Epigenetic Changes in Cancer

Another control system is genomic imprinting, a process whereby one allele at certain genetic loci is transcriptionally silenced in a parent-specific manner, without permanent changes in DNA. In Lecture 3, where we first met genomic imprinting, we used examples from chromosome 15. *Prader-Willi syndrome* arises when the paternal copy of a specific region on chromosome 15 is lost or inactivated, leaving only the inactive maternal allele. *Angelman syndrome* develops when the maternal copy of a nearby region on chromosome 15 is lost or inactivated, leaving only the inactive paternal allele. Imprinting is associated with differences in *DNA methylation* of the two alleles, but the molecular mechanisms are not yet fully understood.

Recent studies have implicated changes in genomic imprinting in several cancers, thus suggesting that epigenetic processes may contribute to tumorigenesis. A prominent example is *Beckwith-Wiedemann syndrome (BWS)*, which is a disorder characterized by prenatal overgrowth and cancer; it affects 1 child in 15,000, mostly on a sporadic basis, but when inherited it behaves as an autosomal dominant trait. Children with BWS are large overall, with noticeable enlargement of the tongue and a variety of craniofacial dysmorphologies. About 20% of them have a tumor that arose prenatally, including *Wilms tumor* (a childhood kidney tumor) and several others. The loss or disruption of a genomic segment at 11p15 from the maternal chromosome is associated with development of BWS, which suggests that an imprinted locus is in that region and the paternal allele is normally silenced.

One of the genes located at 11p15 is *IGF2 (insulin-like growth factor 2)* and another is *H19*, a gene that encodes a nontranslated but abundant RNA of unknown function. Both *IGF2* and *H19* are imprinted in humans and in mice, but *IGF2* is expressed from the paternal allele and *H19* is expressed from the maternal allele. In Wilms tumors, expression of both alleles occurs at the *IGF2* locus 70% of the time, and expression of both alleles at the *H19* locus occurs 30% of the time. Thus, there is frequent *loss of imprinting (LOI)* in Wilms tumor. There are now at least a dozen tumors where loss of imprinting at either the *IGF2* or *H19* locus or both has been demonstrated. Two or more other genes in the 11p15 region also show genomic imprinting and LOI in BWS patients.

Is LOI a cause or an effect of tumorigenesis? We don't have a definitive answer to that question yet, but it is significant that IGF2 protein is a mitogen. Animal studies imply that LOI is essential for tumor progression in certain situations, and LOI implies that the *IGF2* locus is expressed. Another imprinted locus is $p57^{KIP2}$, which is normally expressed only from the maternal locus, but is totally silenced in nearly all Wilms tumors. There are other examples of epigenetic changes in cancer, at loci where imprinting does not occur. One is a gene known as *Apaf-1*, whose activity is essential for the decision to enter apoptosis. Malignant melanomas frequently silence that gene transcriptionally, without mutating it; in cultured melanoma cells it can be reactivated by treatment with an inhibitor of DNA methylation.

One can't help wondering whether changes in the stable expression states of important regulatory genes can "just happen" (that is, as a consequence of random fluctuations), whether environmental agents can alter a pattern like genomic imprinting, or whether there first has to be a mutation somewhere that affects the maintenance system, thus placing epigenetic changes as secondary effects in tumor initiation and progression. We will have to add that question to the list of problems about the molecular basis of cancer that remain to be solved.

REVIEW

Cancer is a class of genetic diseases that result from loss of cell growth control caused by a wide variety of mutations. Development of a tumor typically begins with an initiating mutation that disturbs the control mechanisms for cell multiplication. Expression of abnormal multiplication potential may require promotion—receipt of a mitogenic signal. Progression to full malignancy, which involves the ability to invade adjacent tissues and eventually to metastasize (establish cancer colonies distant from the initial site) generally requires several more mutations. Those secondary mutations are the result of genetic instability, which may arise from loss of control of DNA replication or DNA repair, in ways not yet fully characterized.

Cancer initiating mutations take place in two classes of genes: proto-oncogenes and tumor-suppressor genes. The normal functions of most proto-oncogenes involve stimulation of the cell cycle, under appropriate conditions for cell growth and multiplication. Proto-oncogenes can be converted to phenotypically dominant oncogenes by mutations in the gene itself or in its regulatory regions, by amplification of the copy number, or by chromosomal rearrangements. Tumor-suppressor genes have the general function of inhibiting progression through the cell cycle and/or maintaining control of cell-cell interactions. Tumor-suppressor genes are phenotypically recessive; mutations in both alleles or loss of both normal alleles are necessary before loss of cell growth control results.

Control of the cell cycle is a complex process and the details are not fully known. One major checkpoint is in late G1 phase, where the decision to proceed to DNA replication (S phase) is made. Another important checkpoint is in G2, where the cell decides whether errors in the DNA sequence have been repaired and it is all right to enter mitosis. Mutations in the genes that control DNA repair are often involved in tumor initiation or progression; they can also be the basis for the characteristic genomic instability of cancer cells, which enables them to mutate at a much higher rate than normal cells. Epigenetic changes caused by gain or loss of genomic imprinting can also be involved in development of a malignant cancer.

The lecture described several types of cancer that illustrate one or more of the above generalizations. The examples were colorectal cancer, Li-Fraumeni syndrome, breast cancer, xeroderma pigmentosum, Bloom syndrome, and Werner syndrome.

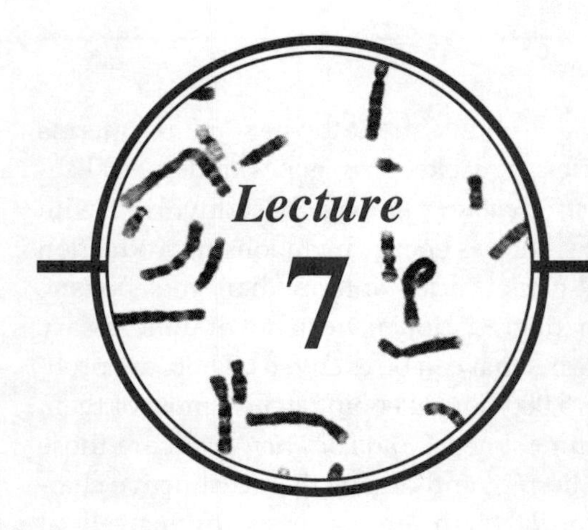

The Human Genome and Developmental Biology

Development of a zygote into a newborn child and the further development of a child into an adult human is a process of mind-boggling complexity. The fact that the outcome of a sperm fertilizing an ovum routinely results in the birth of a normal child is one of life's everyday miracles, even though we know that, in principle, it can be explained in terms of genes interacting with environmental factors. However, we are only in the early stages of a very long quest whose goal is to translate that theoretical generalization into a detailed molecular description of gene activity at all stages of development in all organs and tissues. Analysis of gene expression is particularly difficult in embryonic and fetal humans, because of the difficulty of obtaining suitable material. We have to rely heavily on experiments with transgenic mice and on serendipitous availability of human prenatal material. Nevertheless, significant progress has been made, and the power of modern analytical techniques guarantees that vast amounts of data can be collected from small or rare samples. In this lecture we can do no more than outline some of the major themes in human developmental genetics.

Before we talk about specific genes or genetic disorders, this is a good time to learn the basic facts about construction of transgenic mice, as outlined in Box 7-1. Transgenic mice are essential to the study of gene expression in mammalian development. One can make a mouse with a mutant allele at any genetic locus, either in heterozygous or homozygous state, and ask what happens during prenatal and/or post-natal life of that animal. Transgenes can be incorporated randomly into the host genome (where they sometimes don't behave normally) or, with more effort, transgenes can replace the corresponding gene in the host via homologous recombination. One can replace mouse genes with human genes, thereby making it possible to create mouse models of human genetic diseases. In transgenic mice, one can study the effect of various new drugs that are ultimately intended to treat human diseases. They are especially useful for studying the effect of mutant alleles in the nervous system, which often cannot be approached at all with human material. Of course, we need to remember that humans are not large mice, so some experimental questions about human genes cannot be answered with mouse experiments; nevertheless, a large fraction of what we now know about gene expression during mammalian development (limited though it may be) has been obtained with transgenic mice.

As the first example of a significant problem in human development, we will begin with a paradox. Human reproduction is uniquely and inexplicably inefficient! Only 50–60% of all human conceptuses lead to live births, which is quite different from the situation in lab animals and livestock. For example, if a mouse has ovulated 10 mature ova, she will routinely produce 10 pups. Among the 40–50% of human conceptuses that are lost, about three-fourths fail to implant, and thus are usually undetected as pregnancies. One of the major contributors to this *fetal wastage* is aneuploidy and other chromosomal abnormalities. From Lecture 2 you are familiar with trisomy 21 (Down syndrome), which is the most common aneuploidy found in live births; but the vast majority of all aneuploidies result in fetal loss. As with trisomy 21, the frequency of aneuploidies in general increases with maternal age. We don't yet understand this phenomenon of human fetal wastage, but it is likely to have complex causes, including genetic polymorphisms and environmental factors. Nor do we understand why humans are different from other mammals in this way. Is there positive selection for this aspect of our reproductive biology? To be sure, we don't *need* more efficient reproduction; we have already overpopulated this planet and its ecosystems are in grave danger from depletion of resources and pollution. It would be nice to know why humans have exceptional fetal wastage, but there is no crisis to be averted with that knowledge.

Within the subset of conceptuses that lead to live births, congenital malformations occur at a nontrivial frequency. Congenital malformations are developmental abnormalities that are detectable by visual inspection or routine clinical examination. Approximately 2–3% of living newborns have a congenital malformation, and that number rises to 4–6% when abnormalities diagnosed within the first 2 to 3 years of life are considered. It is estimated that 20% of infant mortality can be attributed to congenital malformations. Both environmental and genetic causes of congenital defects are known, but a large fraction of cases arise from unknown or undiagnosed causes.

Most readers of this book are already familiar with one or two spectacular examples of environmental sources of developmental abnormalities, such as the epidemic of infants with paddle-shaped limbs (*phocomelia*) who were born to mothers taking the sedative, *thalidomide*, several decades ago. *Fetal alcohol syndrome*, which can be caused by heavy consumption of alcohol by pregnant women, is relatively common; it is characterized by mental retardation, heart defects, several mild facial abnormalities, and poor post-natal growth rate. *Cocaine* use during pregnancy can affect fetal brain development, which leads to various behavioral abnormalities after birth. The antibiotic *streptomycin*, taken during pregnancy, can cause deafness in the child. *Folic acid deficiency* during pregnancy can produce *anencephaly* (the absence of all or most of the brain). In some countries, anencephaly tends to be more common among infants born in January, which sounds puzzling at first, but it may be related to nutritional deficiencies that develop during the winter months, culminating in April, when January babies are conceived.

Much more could be said about the influence of environmental factors on prenatal development, but this is a genetics book, so we shall concentrate on the genome. You already know the answers to cosmic questions such as, "Do genes play a major role in development?" Of course they do; development is a saga of gene activation and gene inactivation, varying in time and place. Gene products interact with each other and with metabolites in thousands of reactions, all of which must perform accurately to produce a normal, healthy newborn.

One way to think about the genome is in terms of two categories: housekeeping genes and specialization genes. *Housekeeping genes* are those involved with basic operations, such as energy metabolism, production of proteins and nucleic acids, intermediary metabolism, and catabolism (degradation of worn out or unnecessary molecules to forms that can be excreted). There are probably 10,000 to 15,000 housekeeping genes, most of them being active in most cells. *Specialization genes* are those that give cells, tissues, and organs their distinctive characteristics. Some of them are expressed exclusively at limited times during development; some are expressed first at a particular stage of development, then throughout the life of the organism in a given cell type; and some of them are expressed at several times and places, with interspersed periods of nonexpression. Specialization genes probably represent at least half of all human genes. In this lecture we shall limit our coverage to a few examples of outstanding importance or interest.

PRENATAL DEVELOPMENTAL GENETICS

In humans, the time between conception and birth is traditionally divided into the embryonic period (first eight weeks) and the fetal period (nine weeks to birth). Body form and the rudiments of most organ systems are established during the embryonic period. It is the time when genetic abnormalities usually have the most drastic effects.

Transcription Factors and Early Development

Development is characterized by hierarchies of transcription factors (TFs). When we know the whole gene expression story for development of any given organ or tissue, it will surely contain an intricate network of TFs initiating one stage of specialization, which leads to the activation of other TFs that control the next stage, which turns off some of the first stage TFs and activates others, which feedback in complex ways onto everything that has already happened. Let's begin with a basic example.

A fundamental aspect of early development is differentiation of the embryo along the anterior-posterior axis. In fruit flies, the major anterior-posterior regions are the head, thorax, and abdomen; in humans and other mammals they are the various portions of the brain and the spinal cord. In all multicellular animals, from jellyfish to genetics students, the basic features of axial differentiation are controlled by a group of genes called *HOX*.

The HOX proteins are transcription factors that have in common a motif of about 60 amino acids, the *homeodomain*, which defines their DNA-binding specificity.

HOX genes were first discovered in *Drosophila*, because some HOX mutations are responsible for *homeosis*, the spectacular transformations of one organ into another, such as antennae into legs. In all organisms studied so far, *HOX* genes occur in clusters. Humans (and most vertebrates) have four clusters (A,B,C,D), each with 13 genes. Within a cluster, there can be considerable variation in the amino acids encoded by the different homeodomains or adjacent regions, but the sequences of *paralogs* (genes in the same position in each cluster) are highly similar. Thus, *A10* is more similar to *B10*, *C10*, and *D10* than it is to *A9* or *A11*. This implies that the paralogs have very similar, but not necessarily identical, functions.

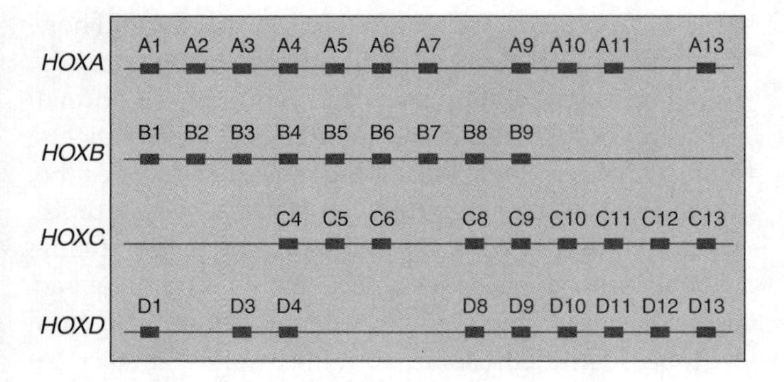

Figure 7-1 The human HOX gene clusters.

A striking fact about the expression of *HOX* gene clusters is that they are expressed in the same order as they are arranged in the genome: Gene 1 is expressed in the most anterior structures, gene 13 is expressed in the most posterior structures. Patterns of *HOX* expression are not sharply delineated; there is considerable overlap from one body region to another, with more *HOX* genes being active in posterior regions than in anterior regions.

What role do *HOX* genes play in human genetic disease? So far, only a few *HOX* mutations have been identified, including *HOXD13*, which can cause *synpolydactyly* (more than five digits, fused) and *HOXA13*, which can cause *hand-foot-uterus syndrome* (abnormal thumbs, small feet, duplication of female genital tract). This should not surprise you. Any mutation that interferes significantly with the function of a transcription factor that is crucial for early embryogenesis will be lethal, probably even before pregnancy is evident. Only mutations with relatively mild effects on development will become available for analysis. In that connection, we should realize that *HOX* genes, like many other transcription factors that are important for early developmental processes, also function later in development in various specific organs or tissues. Mutations that affect those later functions are more likely to come to the attention of geneticists. An example is provided by *HOXA10* and *HOXA11*, where mutations are known that affect uterine receptivity—an adult function.

HOX genes and their encoded proteins illustrate some basic facts about gene expression during development: complexity and interactions. It would be a mistake to think in simplistic terms such as, "Gene *XYZ* makes protein XYZ, which binds to control sequences at 10–20 other genes and turns them on or off." Reality is far more complex. Gene *XYZ* may be transcribed at different rates in different parts of the embryo, and it may be subject to alternative transcriptional forms, so that it encodes a family of proteins. The proteins produced from gene *XYZ* almost certainly interact with several other transcription factors to determine whether target genes are activated or repressed, so the overall effect of the products of gene *XYZ* may vary widely. We are in the early stages of understanding transcriptional networks, especially during the prenatal stages of mammalian development. We do not know how many genes are the targets of a specific HOX protein, we do not know how many other transcription factors act in concert with a specific HOX protein, nor do we know the identity of the molecules (small or large) that affect the transcription of *HOX* genes.

PAX proteins are another group of transcription factors that are important in early development, where they act as regulators of *organogenesis*. They are also essential for maintaining *pluripotency* of stem cell populations, the ability to differentiate into several specialized cell types. PAX proteins apparently bind to enhancer DNA sequences, thereby modifying transcriptional activity of downstream genes. A definitive aspect of PAX proteins is the presence of a 128 amino acid *paired domain*, which forms two DNA-binding regions. Humans (and mammals in general) have nine *PAX* genes, most of which have been implicated in developmental disorders when mutated.

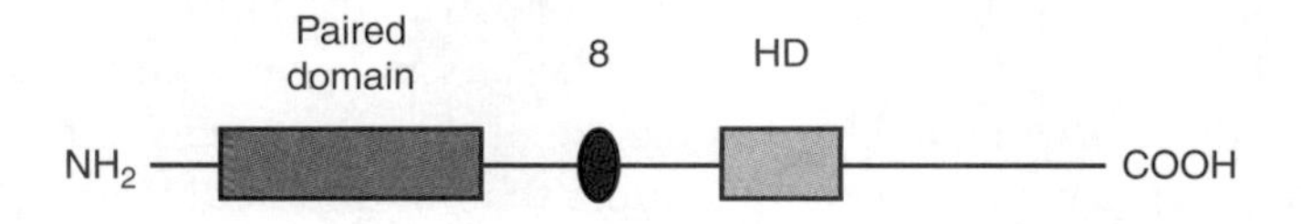

Figure 7-2 A *PAX* gene. Paired domain—see text; 8 = a region found in most *PAX* genes that encodes an octapeptide; HD = homeodomain.

PAX1 and *PAX9* are expressed in the developing vertebral column, the limb buds, and the thymus, where they have overlapping expression patterns. In mice, mutations in *PAX1* cause axial skeleton defects, whereas mutations in *PAX9* can lead to absence of the thymus and parathyroid gland, as well as a variety of skeletal defects. In humans, a mutation in *PAX9* causes *oligodontia*—failure of six or more of the permanent teeth to develop.

PAX3 is expressed early in mammalian development in the dorsal neural tube, the region from which the migratory neural crest cells arise. They contribute to a variety of organs, including the heart, the peripheral and enteric ganglia, melanocytes, and Schwann cells. Enteric ganglia defects cause gastrointestinal motility disorders, such as *Hirschsprung disease*. *PAX3* mutations also cause one form of *Waardenburg syndrome* in humans, probably because normal neural crest cells are needed in the middle ear for hearing. *PAX6* is crucial for ocular development. It is expressed in the developing optic cup and the overlying ectoderm from which the lens will form. In humans, heterozygous *PAX6* mutations lead to *aniridia* (absence of the iris) and *cataracts*, among other defects.

The functions of PAX proteins are complex. They tend to be expressed at more than one stage of development; they interact in various ways with other transcription factors; and they are subject to alternative splicing, which affects their functions in a tissue-specific manner. We can confidently expect that many more human developmental disorders will be shown to result from mutations in *PAX* genes.

Organogenesis

The thyroid gland offers an example of the many ways in which genes can affect the development and function of an organ. In Lecture 4 you learned that one of the most common birth defects is *congenital hypothyroidism (CH)*, occurring once in every 3,000 to 4,000 births. We used CH as an example of successful population-wide genetic screening, because correction of the abnormality is effective and cheap, requiring nothing more than daily intake of synthetic thyroid hormone in a pill. It turns out that CH can be caused by mutations in several genes whose expression is important at various stages of development—a generalization that undoubtedly applies widely to genetic disorders of organogenesis.

Approximately 85% of CH cases result from *thyroid dysgenesis*; that is, the gland is either much smaller than normal, absent altogether, or in an abnormal location. During embryonic development, the thyroid gland begins as an endodermal bud in the posterior region of the pharynx floor. It migrates dorsally and caudally until it reaches the anterior wall of the trachea (at about 8 weeks of gestational age in humans). Then some cells begin to differentiate into *thyroid follicular cells (TFCs)*, where thyroid hormone will be made. This requires a variety of organ-specific genes, including thyroglobulin, thyroperoxidase, and the receptor for thyroid stimulating hormone (TSH). There is extensive proliferation of TFCs for the next several weeks. Three transcription factors—TTF1, TTF2, and PAX8—are among the regulatory proteins required for thyroid gland development and differentiation. Mutations in each of them have been shown to be responsible for some cases of congenital hypothyroidism.

Another factor required for normal thyroid hormone is the *TSH receptor*. In addition to being part of a signaling cascade, its presence is required for normal proliferation of thyroid follicular cells (TFCs). Patients with two abnormal alleles of the gene that encodes the TSH receptor display thyroid *hypoplasia*, as well as unresponsiveness to TSH. In addition, there are many patients with thyroid dysgenesis for whom no genetic defect has been identified. Presumably additional genes will be identified, some of which may interact to produce complex patterns of inheritance that lead to the CH phenotype.

In a normal thyroid, TSH is synthesized by the pituitary, in response to *thyrotropin releasing hormone* from the hypothalamus (thyrotropin is another name for TSH), see Figure 7-3. When it binds to the TSH receptor on the surface of TFCs it stimulates the production and secretion of *thyroid hormone*. Interruptions in that process are called *dyshormonogenesis*; they account for 10–15% of congenital hypothyroidism cases. A central element in thyroid hormone synthesis is *thyroglobulin*, a large dimeric protein encoded by a gene that spans more than 300 kb and contains 48 exons. It is modified by iodination on several tyrosine residues, secreted, taken up again by TFCs and broken down in lysosomes, leading to the iodinated tyrosine derivatives called T4 and T3, which are the *thyroid hormones*. The thyroid gland produces considerably more T4 than T3. Both are secreted into the blood, where they regulate energy metabolism in virtually all tissues. Most T4 is converted to T3 in the liver, and T3 is actually the more active hormone, because it binds more efficiently than T4 to the hormone receptor in cell nuclei.

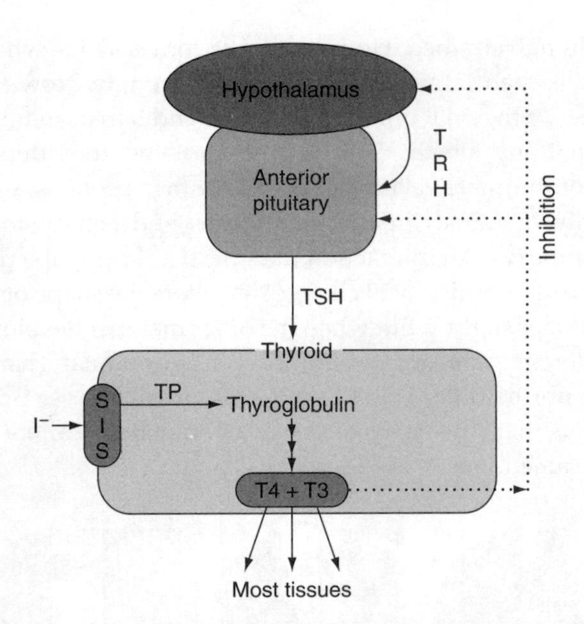

Figure 7-3 Thyroid hormone regulation. TRH = thyrotropin releasing hormone; TSH = thyroid stimulating hormone; SIS = sodium/iodide symporter, TP = thyroperoxidase.

Patients with defects in the gene for thyroglobulin are hypothyroid and usually have *goiter* (an enlarged thyroid gland). Another common source of failure to synthesize thyroid hormone is the gene for *thyroperoxidase*, the enzyme that adds iodine to tyrosine residues in thyroglobulin. Other CH patients have mutations in the *sodium/iodide symporter*, the membrane protein required for uptake of iodine into thyroid cells. And so it goes: Anything that can go wrong will go wrong, at least in some unlucky people. Murphy's Law is alive and vigorous during the development of the thyroid gland, and there is good reason to believe that similar complexities affect the development of every other organ in the human body.

As a second example of genetic complexity underlying organogenesis we will look at one aspect of limb development. Although the genetic systems are only partially understood, detailed analysis of limb development does not seem to be beyond reach in the foreseeable future. *Limbs have three axes of symmetry: proximodistal, anteroposterior, and dorsoventral.* For each axis, specific genes are crucial, but extensive overlap and interaction exists among the sets of genes required for normal development of all three axes. For the proximodistal axis several *fibroblast growth factors (FGFs)* and their receptors play important roles. Mutations can lead to various anomalies, including syndactyly (fused digits), brachydactyly (short terminal phalanges), fused or curved long bones, and so on.

One of the most striking anomalies of proximodistal limb development is *achondroplasia (ACH)*, an autosomal dominant disorder that is the most familiar form of dwarfism. It is characterized by marked shortening of the proximal segments of arms and legs, combined with frontal bossing, lumbar *lordosis* (forward curvature of the spine), limitation of elbow extension, and several other minor anatomical effects. Mutations in the gene for fibroblast growth factor receptor 3 *(FGFR3)* are responsible for ACH. Remarkably, virtually all the mutations in *FGFR3* that cause ACH are in one place, producing a substitution of arginine for glycine at position 380 (G380R), which is in the transmembrane domain of the protein. Nearly all of those mutations are at position 1138 in the nucleotide sequence, which makes that spot one of the most mutable sites known in the human genome—at least 1,000 times the background mutation rate. The reason for the unusual instability of that site is completely mysterious.

The consequences of the G380R substitution are also unclear. Adding to the confusion is the fact that mutations at other points in the *FGFR3* gene do not cause achondroplasia. Another highly mutable site at amino acid position 250 is involved in *Muenke nonsyndromic coronal craniosynostosis* (a type of cranial bone fusion), whereas mutations at some other sites in *FGFR3* cause *thanatophoric dysplasia* (a lethal neonatal dwarfism). In principle, the different phenotypes produced by mutations at different places in one gene can be explained by interactions between the product of that gene and various other proteins, so it may be that FGFR3 protein can transmit different messages to the interior of cells, depending on what other proteins associate with it. For now, we have to be satisfied with speculation.

Complex Inheritance of Developmental Abnormalities

In Lecture 3 we discussed the role of multiple genetic factors in causing human diseases such as diabetes, cardiovascular disease, and mental illness. Developmental abnormalities are especially prone to originate from the cumulative effects of variant alleles in more than one gene. This is to be expected, because dozens or hundreds of genes may be crucial for the formation of a normal organ, such as the eye, the ear, the liver, or the genitalia. Mutations that completely abolish the function of one essential gene in a developmental sequence are often *embryonic lethals*, depending on the structure involved. Other single gene mutations may be compatible with life, but lead to single-organ defects such as deafness or blindness, with classic Mendelian inheritance. However, a growing number of developmental abnormalities appear to be caused by *digenic inheritance*, the cumulative effects of mutations in more than one gene.

Holoprosencephaly (HPE) is a developmental abnormality characterized by incomplete separation of the forebrain into distinct left and right halves. It is fairly common prenatally, occurring in about 1 in 250 conceptuses, but only 1 in 16,000 live births. Affected individuals usually display mental retardation and various facial abnormalities, such as cleft lip and palate, cyclopia (one centrally located eye), microcephaly, or one maxillary incisor. In familial HPE, there is great variability in symptoms among obligate carriers, from severe to mild to no clinical abnormality. This is presumptive evidence for the existence of modifier genes.

At least eight genes have been shown to cause HPE, but they account for less than one-fourth of all HPE cases that have been analyzed, so there must be more genes capable of causing the condition. The gene most commonly associated with HPE so far is *sonic hedgehog (SHH)*, a cell-cell signalling gene involved in several developmental processes; but in some families, a mutation in one copy of SHH is not sufficient to produce HPE. Analysis of several families has shown that a mutation or deletion of another gene, in combination with the SHH mutation, is necessary for the HPE phenotype.

In principle, digenic inheritance can have several origins, each involving genetic variations that are not individually severe enough to disrupt normal functions. One possibility is that two or more steps in a metabolic sequence operate at less than normal efficiency. Then, the overall sequence may be slowed to an extent that is not compatible with normality. Another possibility, which is especially relevant to development, is that two or more stages in a sequence of gene expression patterns may be subnormal. Thus, a structure formed at one stage may be slightly abnormal, and this affects what it can become at the next stage. If the next stage is also slightly abnormal, the cumulative effect can be a defective organ when the infant is born.

A third explanation for digenic inheritance, mentioned in Lecture 3, is that mild alterations of the shape of two or more interacting proteins may add up to no interaction at all. Human relationships provide an analogy. Suppose individuals A (a man) and B (a woman) have an intimate relationship. After a while, A meets woman C, who appeals to him very much, but he is happy with B, so the A+B bond remains functional, although strained. However, there comes a time when B meets man D, who appeals to her strongly. Now both sides of the A+B union are stressed and it just isn't functional any longer. So it is with proteins; they depend upon intimate relationships with other proteins, especially in transcriptional complexes and cell signaling complexes. An interaction may persist despite the presence of an amino acid change that alters the shape of one member slightly, but when the other member develops a different, noncomplementary configurational change, the normal interaction becomes impossible or so inefficient that the normal level of function cannot be sustained.

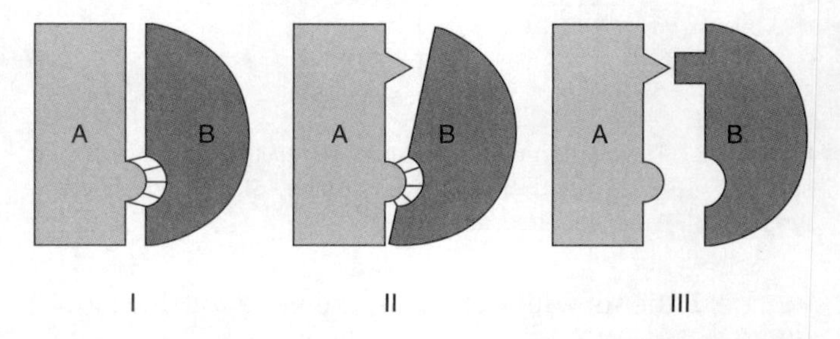

Figure 7-4 Protein relationships. Panel I—the normal interaction between A and B; panel II—A has changed in a way that strains the relationship, but does not abolish it; panel III—both A and B have changes (caused by mutations in the corresponding genes) and normal interaction is no longer possible.

Abnormalities in left-right asymmetry offer another example of genetic complexity. Although we are fundamentally *bilaterally symmetrical* (two eyes, ears, arms, legs, etc.), many other anatomical features are neither paired nor located on the midline. Familiar examples are the heart and stomach (left side) and the liver (mostly right side). The lungs, although paired, are unequal, with three lobes on the right and two lobes on the left. Most organs with asymmetric features begin development in the midline, then *lateralize* (move to one side). An alternative process applies to some major arteries and veins, which begin as paired structures, followed by regression of one member.

Approximately 1 in 5,000 live births has mirror-image organ asymmetry, either of a single organ (such as the heart) or all of them. The term *situs solitus* refers to the normal pattern of asymmetry and *situs inversus* (or *situs ambiguous*) is used for either complete reversal of asymmetry or partial reversal. Frequently, there are no functional consequences, but sometimes the rearranged organ placement is also associated with structural and/or functional abnormalities.

Genetic studies on mice have already implicated at least 20 genes in the etiology of left-right reversal. Among them are sonic hedgehog (mentioned in the preceding section), various members of the *TGF-beta* signaling family, an X-linked transcription factor (ZIC3), and several motor proteins that also are required for motile cilia. It is not yet possible to put all the steps into a logical and coherent causation map. Moreover, there is evidence that digenic inheritance may be involved here, because some mice heterozygous for mutations in either one of two genes are normal, but mice that are double heterozygotes have situs inversus or related abnormalities. The implications for clinicians who want to know why a particular patient with a left-right asymmetry problem is abnormal are not encouraging.

Genomic Disorders

In Lecture 2 you learned that a common cause of human genetic disease is duplications and deletions of large genomic segments; several examples affecting adult functions were presented. Genomic disorders, especially those involving several genes (*contiguous gene syndromes*), often affect developmental processes, so that a child with multiple congenital abnormalities is born. Following are some well-known examples.

Velocardiofacial syndrome/DiGeorge syndrome (VCFS/DGS) is usually associated with a deletion of about 3 Mb from 22q11. Its frequency is 1/4,000 live births. It causes abnormalities in facial appearance, the pharynx, speech, palate, and heart, plus learning disabilities. On either side of the deletion-prone region, there is a 200 kb repeated segment. Unequal recombination arising from misalignment of portions of those repeated segments causes the VCFS/DGS deletion. Although a transcription factor gene called *TBX1* may be responsible for most of the clinical findings, the contribution of the other genes in the deleted regions remains to be elucidated.

Cat-eye syndrome is a rare genomic disorder associated with tetrasomy (four copies) of a portion of chromosome 22, including the p arm and adjacent 22q11. Clinical features are complex, involving the eyes, heart, kidneys, face, genitals, and skeleton, as well as mental retardation. Candidate genes are under investigation.

Smith-Magenis syndrome (SMS) is characterized by multiple abnormalities, including brachycephaly, proganthism, hoarse voice, mental retardation, and behavioral problems, among which are self-mutilation. Most patients have an approximately 5 Mb deletion of 17p11.2, which arises because of unequal recombination between 200 kb repeats that flank the deletion region. As you will recall from Lecture 2, when unequal recombination produces a deletion on one chromosome, the reciprocal product is a chromosome with a corresponding duplication. It is gratifying to learn that a few patients with mild retardation and minor *dysmorphic* features have recently been shown to possess a duplication of the same region deleted in SMS patients. Here we have another example of the general rule that the phenotypic consequences of having three copies of a few genes (trisomy) are usually milder than the consequences of monosomy for the same genes.

GENETICS OF AGING

Development does not cease at birth. We spend two decades acquiring mature form and function, and shortly thereafter various metabolic processes enter a long process of decline, changing slowing at first, but with increasing speed after the fifth or sixth decade of life. Aging in humans is universally recognized to be associated with decreases in lung capacity, muscular strength, cardiovascular function, and reproductive capacity. Simultaneously, aging is accompanied by an increased probability of life-threatening diseases, such as cancer, stroke, and heart attack.

What is the role of the genome in aging? One way to approach the genetics of aging is to measure the *heritability of life span*. Studies on twins and the Amish (an inbred population) indicate low heritability—in the range of 30%. However, few individuals in those cohorts lived beyond 85 years. A different perspective emerges from studies on *centenarians*, whose siblings also tend to have exceptional life spans. For example, siblings of centenarians in one study had a fourfold greater probability of surviving to greater than 85 years than siblings of persons who died by age 73. One recent study on centenarians and their close relatives has implicated a gene on chromosome 4 as having a role in attaining extreme old age. Based on the limited heritability data so far available, one would conclude that environmental factors are more important than inheritance in determining life span for most people, but there probably are a few genes that significantly affect longevity.

How might the genome affect longevity? Several possibilities must be considered. First, the gradual accumulation of mutations throughout the genome may lead to a nonspecific decline of function in all parts of the body. This seems unavoidable, but whether it is the most important factor in limiting life span cannot yet be evaluated. Second, there may be specific genes that accelerate aging when mutated. Common sense tells us that any impairment in the system for DNA repair or any mutation that would increase the frequency of mutations could lead to more rapid aging, as well as an increase in the frequency of cancer and other metabolic diseases. We mentioned Werner syndrome in Lecture 6 on cancer, and in Lecture 5 we considered the possibility that abnormalities of mitochondrial function might influence aging.

Third, there may be specific genetic variants that diminish the rate of aging. Very little is known about this topic for humans, but studies on several model organisms have identified a few genes where inactivating mutations significantly increase lifespan. Examples are the genes for growth hormone (GH) and its receptor, the gene for insulin-like growth factor 1 (IGF-1), and the gene that encodes SHC, a histone deacetylase. Reductions in GH and/or IGF-1 have widespread effects on metabolic rate, and deficiencies in SHC activity reduce the rate of transcription. These effects correlate with the long-standing observation that caloric restriction will extend the life span of organisms from worms to mice.

A major underlying mechanism of longevity enhancement, whether by caloric restriction or mutations in specific genes, appears to be generation of fewer reactive oxygen species (ROS). As you learned in Lecture 5, ROS of various types are produced as a byproduct of electron transport in the mitochondria primarily (although some other reactions also produce ROS in the cytoplasm), and ROS are a major source of mutations. When the rate of ROS production slows, fewer mutations will occur for two reasons: first, the enzymes that convert ROS to harmless products can do their job more thoroughly, and second, the DNA repair enzymes have more time to correct mutations before they are established and passed on to new cells.

Obviously, there is intense interest in finding out whether we might be able to slow the aging process, especially if it can be done by extending the period of middle life, where there is still vigor and relatively little degenerative disease. Applying the information obtained from model organisms to humans will not be simple, however. For example, although growth hormone deficiency extends the life span of mice, it also produces dwarfism. GH deficiency in humans is known to be associated with many problems, including hypertension, insulin resistance, and atherosclerosis. The challenge for aging research will be to find metabolic regimens that extend life span without seriously compromising the quality of life. Efforts will be made to find ways of maintaining a normal metabolic rate without generating the usual quantity of ROS. Another approach will be to enhance the efficiency of the DNA repair system, which would not only have a general effect on the accumulation of somatic mutations, but also would reduce the frequency of cancer.

A LOOK AHEAD

This lecture ends with a preview of complexities to come. At the end of Lecture 1, you learned about a new class of RNA involved in translation control. The phenomenon is called *RNA interference*, referring to the fact that these small RNAs interfere with expression of specific genes at some level. The RNAs go by several names: small interfering (siRNA), short temporal (stRNA), and micro (miRNA); we will use *siRNA* here. Discovered in plants, RNAi has now been extensively documented in worms and fruit flies and has been shown to exist in mammals, including humans. RNA interference is going to have a significant impact on the study of developmental gene regulation.

The basic idea is that the genome contains short nucleotide sequences that are complementary to a portion of the mRNA sequence of specific genes. By mechanisms yet to be documented in detail, siRNA genomic sequences are transcribed into double-stranded RNA (dsRNA), then processed to smaller pieces. The products, which are in the range of 21—26 nucleotides, either become single stranded, bind to their target mRNAs and shut down translation, or act as guides for an enzyme complex that cleaves target mRNAs, rendering them nonfunctional. This was illustrated in Figure 1-17.

It is already clear that there are more than 250 siRNA genes in humans, but the total number is unknown. A popular hypothesis is that RNA interference evolved as a defense against viruses, many of which have a double-stranded RNA stage in their life history. In addition, RNAi may protect against damage from transposons and retrotransposons, but there is no proof of either possibility as yet. Data from the worm and fruitfly strongly suggest that RNAi is also involved with developmental gene regulation, because mutations in specific siRNA genes produce abnormalities at specific times and places. Whatever the normal function of siRNA may be, this class of molecules has rapidly been developed into a powerful experimental tool by biologists working with the mouse and other experimental organisms. For example, it is relatively straightforward to prepare a synthetic dsRNA that contains a sequence complementary to the mRNA of a gene of interest, introduce that dsRNA into cultured cells by transfection, and then measure the expression of the targeted gene and/or general effects on gene expression in those cells. For some purposes, the transient nature of these "knockdown" experiments is more suitable than the permanent effects produced by structural inactivation of a gene. As the technology becomes more sophisticated, it will undoubtedly be used productively in studies of developmental gene regulation in animal models. Furthermore, we can anticipate that a new class of human developmental anomalies will be identified, resulting from the effects of mutations in siRNA genes.

Finally, let's consider what the future of human developmental genetics will be. How much remains to be learned before we fully understand the role of the human genome in development? The answer is, "almost everything." Our knowledge of gene expression during human development is rudimentary. The major obstacle is the scarcity of prenatal human material, despite the large number of pregnancies that are voluntarily terminated every day around the world. This problem will presumably be solved eventually, despite the sensitive ethical and social concerns that accompany it. Microarray analysis will be applied in great detail to human embryonic and fetal material, with an emphasis on detection of alternative transcripts in many cell types. There will also be extensive proteomics studies, aimed at detection of post-translational protein modifications. Analysis of gene expression in the developing nervous system will be particularly difficult, because of the huge number of different cell types, but it will surely be pursued with vigor and determination for many years

Another significant obstacle to the study of human developmental genetics originates from two facts: (1) description of patterns of gene expression does not necessarily identify functional consequences, and (2) one cannot do experiments on humans. This implies that work on nonhuman mammals will be very important in elucidating the genetic basis of human congenital malformations. Studies on transgenic mice will continue to be important, but we cannot learn everything we want to know about humans from mouse experiments. Therefore, there will be a growing need for analysis of gene expression during prenatal development of primates, probably beginning with the rhesus macaque, the most widely studied monkey. Whether full understanding of human developmental genetics will eventually require experimental work with our closest relatives, chimpanzees and gorillas, is a question that cannot be answered now. Certainly the ethical and practical difficulties associated with research on apes are formidable, but who knows what the crucial experimental questions will be for investigators several decades from now?

BOX 7-1
CONSTRUCTION OF TRANSGENIC MICE.

The following figure shows the steps involved in making a mouse where an imaginary wild-type gene for tail formation has been replaced by a recessive allele that does not support tail formation. This is a "knock-out" mouse. The initial construct is made by transfecting ES cells with a two-part transgene: One part confers antibiotic resistance (A^r), so that cells that have incorporated the transgene can be distinguished from those that don't have it; the other part is the recessive allele for taillessness. Each clone of antibiotic-resistant ES cells has to be examined for the presence of the transgene and its location in the host genome. Most transgenes are integrated at random locations, but in this experiment, the goal is to obtain a clone of ES cells that have replaced the host gene with the transgene by *homologous recombination*. Under optimum conditions, we may find 1 in 20 clones of A^r cells that represent homologous recombination. Some of those cells will be injected into a host blastocyst as shown, thereby making a chimeric embryo. As the diagram indicates, production of a mouse that is homozygous for the transgene is a multi-step process.

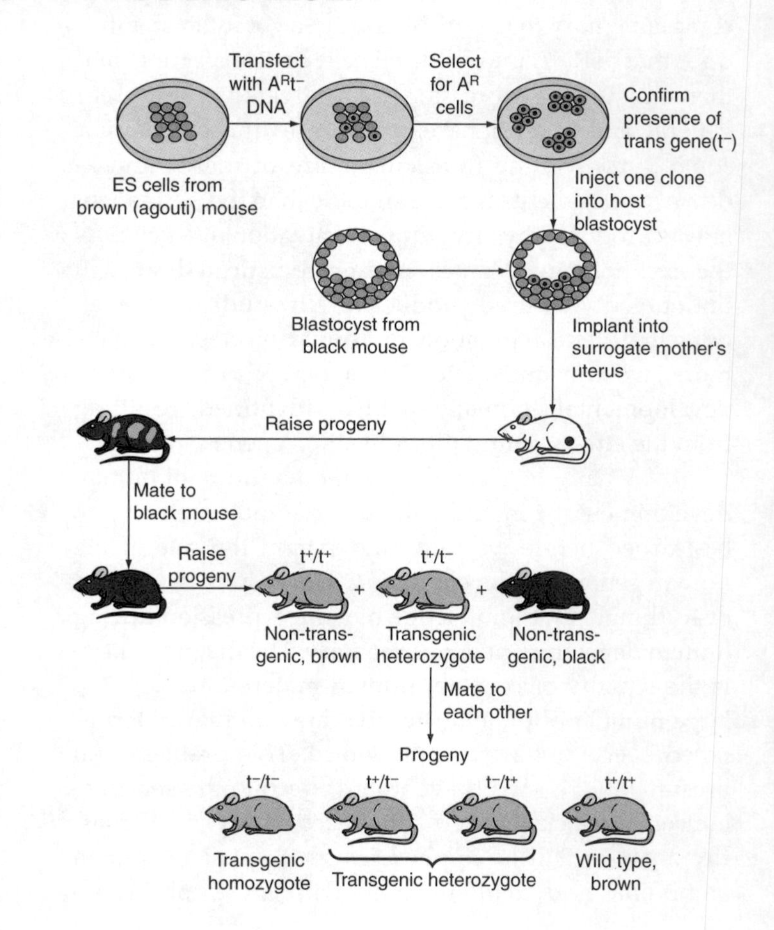

Many variations on this theme are available. For example, the transgene could have been derived from a human, in order to find out how a human allele of medical interest is expressed at various development stages or in response to various drugs. We would call those animals "knock-in" mice. Or, we might have been interested in finding out how a dominant, disease-causing allele from humans affects the functions of normal genes in a mouse; in which case, we might be satisfied with random insertion of the transgene into the host genome. We could also create a transgene that will be expressed only in certain tissues, and we could construct a transgenic mouse in which the transgene can be activated (or deactivated) in response to an external signal, such as a drug.

REVIEW

Molecular analysis of human embryogenesis and development is in its infancy.

Because so little human prenatal material is available, most of what has been learned so far was derived from studies on mice, especially transgenic mice.

Approximately half of all human conceptuses do not lead to live births. Congenital malformations occur in 2–3% of living newborns; some represent genetic disorders, others are produced by exposure of the fetus to various toxins, and many are caused by unknown factors.

Complex patterns of transcription factors, which activate and deactivate large groups of genes at specific times and in specific tissues, control embryonic development. Prominent examples are the evolutionarily conserved HOX genes, which control axial differentiation in all animals, and the PAX genes, which play a major role in organogenesis. A few human genetic disorders can be attributed to HOX or PAX genes, but most mutations in such basic control genes lead to embryonic lethality.

Congenital hypothyroidism was presented as an example of the many ways in which mutations can disrupt development and/or function of an organ, leading in this case to a common (and very treatable) clinical syndrome in newborns. A different class of developmental abnormality is achondroplasia (classic dwarfism), which typically arises from a missense mutation in one specific codon in the large fibroblast growth factor receptor 3 gene.

Developmental abnormalities are often the cumulative result of mutations in more than one gene (digenic inheritance). Examples given were holoprosencephaly and left-right reversal. Deletions and/or chromosomal rearrangements that affect more than one contiguous gene (genomic disorders) are another common cause of complex anatomical and behavioral abnormalities in newborns.

Development does not cease at birth; it continues throughout life, and aging is a developmental process. Extensive work on the effect of genes on aging is underway in model organisms, but little is known about the process in humans. Decline of mitochondrial function may be an important factor.

Understanding the roles of genes and environmental factors in producing human fetal loss and congenital malformations will require a major increase in support for analysis of gene expression during prenatal life.

Glossary

Adenine – One of the four bases found in DNA. A purine, adenine (A) always pairs with thymine (T) on the opposite strand of DNA.

Adenovirus – A medium-sized DNA virus that does not integrate into the host genome. Used commonly as a gene transfer vector.

Allele – Any of the alternative forms of a specific gene.

Alu sequence – A short segment of DNA bounded by sites at which a specific enzyme cuts.

Amino acid – A class of small organic molecules containing an amino group and a carboxyl group. Proteins are assembled from 20 types of amino acids.

Anaphase – The stage of mitosis at which the chromosomes are pulled apart and begin to move toward opposite ends of the spindle.

Aneuploidy – Any deviation from the normal (euploid) number of chromosomes.

Anticipation – The phenomenon of successive generations of persons being affected with a genetic disease at progressively earlier ages.

Apoptosis – Programmed cell death. It may occur spontaneously (as in several phases of development) or in response to mutations or chemical treatment (as in cancer).

Assortative mating – Nonrandom mate selection.

Autosome – Any chromosome other than the sex chromosomes.

Barr body – Microscopically detectable dark-staining region in interphase nuclei representing the inactive X chromosome.

Base pair – One subunit of a double-stranded DNA sequence; either A-T or G-C.

Blastocyst – An early embryonic stage, consisting of several hundred cells in the form of a ball with a cavity.

Cancer – An abnormal growth, capable of invading adjacent tissue and establishing colonies (metastases) in other parts of the body.

Centimorgan – A unit of recombination, referring to a region of the genome where the probability of crossing-over is 1%.

Centromere – A region of repetitive DNA and associated proteins where the meiotic and mitotic spindle fibers attach to a chromosome.

Chromatid – One half of a chromosome after the DNA has been duplicated.

Cis – A term used to indicate a region of DNA near a gene and on the same chromosome (e.g., cis-regulatory sequences). Also used to mean that two specific alleles in neighboring genes are on the same chromosome.

Citric acid cycle – A metabolic pathway that is of central importance in energy metabolism, producing NADH, FADH2, and ultimately, most of the ATP used by cells. Also called the tricarboxylic acid cycle or Krebs cycle.

Codon – A set of three adjacent base pairs in DNA or RNA that specifies an amino acid.

Concordance – Similar phenotype (affected or not affected) in twins, sibling pairs, or other comparisons.

Congenital – A condition that develops during embryogenesis or fetal life; not necessarily genetic in origin.

Cristae – The infoldings of the inner mitochondrial membrane.

Crossing-over – Exchange of DNA segments between paired chromosomes during meiosis, resulting in recombination of linked genes.

Cytosine – One the four bases found in DNA. A pyrimidine, cytosine normally pairs with guanine.

Deoxyribose – The five-carbon sugar found in DNA nucleotides.

Diploid – Two sets of chromosomes; the normal condition in somatic cells.

Dominant allele – A form of a gene that will produce a phenotype regardless of the presence of a normal allele on the homologous chromosome.

Dominant negative – A mutation that causes an abnormal phenotype even though a normal allele is present on the homologous chromosome.

Dosage compensation – The process by which most genes on one X chromosome are inactivated in females, so that equal amounts of their products are produced by both male and female cells.

Dysgenic – Any factor that leads to an increase of disease-associated alleles in the population.

Enhancer – A regulatory region of DNA to which proteins bind that increase the rate of transcription of a nearby gene.

ES cells – Embryonic stem cells, obtained from blastocysts, capable of participating in differentiation of all or almost all tissue types.

Exon – A contiguous segment of DNA that is retained in mRNA after introns have been removed. Most, but not all, exons encode a series of amino acids.

Frameshift – A type of mutation that changes the amino acids specified by the entire region of DNA downstream of the mutation. Usually a deletion or insertion of a number of base pairs not divisible by three.

Free radical – A molecule with an unpaired electron, highly reactive, usually capable of producing mutations.

Gain-of-function – A mutation that creates a new function (which may be deleterious) for a gene product.

Genetic screening – The process of testing members of a population or specific group for genetic variants associated with a disease.

Genome – The total DNA of a species including genes and nongenic sequences.

Genomic imprinting – Differential inactivation of gene copies at a specific locus, depending on which parent the gametes are from.

Genotype – The genetic constitution of an individual, usually with reference to one or a few genes.

Giemsa bands – Colored bands formed on condensed chromosomes with Giemsa stain. Also called G-bands.

Gigabase – One billion base pairs.

Guanine – One of the four bases found in DNA or RNA. A purine, guanine normally pairs with cytosine.

Guthrie Test – A blood test that uses growth of a special bacterial strain to detect children who will develop phenylketonuria.

Haploid – The number of chromosomes found in gametes; half the number in somatic cells. Also used to refer to DNA amount.

Haploinsufficiency – The presence of one normal allele at a locus is not sufficient for normal function or structure.

Haplotype – The set of alleles in a group of linked genes present on one chromosome.

Helicase – A protein that breaks hydrogen bonds between base pairs and separates the DNA strands locally.

Heterochromatin – A region with a large array of tandemly reiterated sequences, usually tightly condensed and transcriptionally inactive.

Heteroplasmy – Presence of a mixture of normal and mutant mitochondrial DNAs in an individual.

Heterozygote – An individual with two different alleles at a specific genetic locus.

Histone – A very basic (positively charged) protein that associates with DNA.

Homozygote – An individual with two normal or two mutant alleles of a specific type at a given genetic locus.

Hybrid cell – A cell with chromosomes derived from two types of parental cells, usually made by cell fusion techniques.

Hydrogen bond – A weak chemical bond between two negatively charged atoms that share a proton (a positively charged hydrogen atom).

Ideogram – A diagram showing metaphase chromo somes with bands made by Giemsa stain or other dyes.

In vitro fertilization (IVF) – Creation of a zygote in a lab, outside of the human body.

Incomplete penetrance – A dominant allele that does not have a phenotypic effect in every individual who carries it.

Intron – A segment of DNA or RNA that lies between exons and does not appear in mRNA.

Inversion – A 180-degree realignment of a portion of a chromosome. If the two breaks were on opposite sides of the centromere, the inversion is pericentric; if both breaks were in one chromosome arm, the inversion is paracentric.

Karyotype – The chromosome complement of an or ganism or cell.

Kilobase – One thousand base pairs.

L1 sequence – The most abundant member of the LINE class of dispersed repetitive sequences.

Ligase – An enzyme that splices two single-stranded segments of DNA together.

Linkage disequilibrium – Nonrandom association of a given allele with one or more nearby polymorphisms.

Lod score – A term used in linkage analysis, indicating the probability that two genetic loci are linked at a specified recombination distance.

Long Interspersed Element (LINE) – A type of reiterated DNA.

Loss of heterozygosity (LOH) – Inactivation or loss of the remaining normal allele in a cell that already has one nonfunctional allele at that locus.

Lyonization – Random inactivation of one X chromo some in every somatic cell of a female.

Lysosome – A subcellular organelle where many macromolecules are broken down.

Macrophage – A type of white blood cell that processes antigens in an early step in stimulating an immune response.

Maternal inheritance – The pattern of inheritance of mitochondrial DNA, which comes exclusively from the female parent.

Megabase – One million base pairs.

Meiosis – The final two cell divisions in gamete formation, during which the diploid chromosome number is reduced to the haploid number.

Mendel's laws – The fundamental rules of transmission genetics, formulated by Gregor Mendel.

Messenger RNA – The type of RNA molecule that is translated into polypeptides on ribosomes.

Metaphase – The phase of mitosis when highly condensed chromosomes line up in the middle of the spindle.

Methylation – Addition of a CH_3 group to a molecule (e.g., to the #5 carbon of cytosine).

Microarray – A very large set of DNA oligonucleotides or cDNAs attached to an inert support in such a way that they can form base pairs (hybridize) with complementary DNAs or RNAs.

Missense mutation – A change in one base pair resulting in a codon that specifies a different amino acid from the original one.

Mitosis – The phase of the cell cycle during which chromosomes condense, attach to the spindle, are pulled apart into two groups, and are distributed to daughter cells so that they have identical sets of chromosomes.

Mutation – Any nonrepaired change in DNA sequence, not necessarily having an effect on phenotype.

Neoplasm – A new growth; a group or mass of cells arising in an abnormal location.

Nondisjunction – Failure of a pair of chromatids to be separated at mitosis, resulting in one daughter cell with three copies of that chromosome and one with only one copy.

Nonsense mutation – A change in base sequence that creates a stop codon in a translated sequence.

Nonsynonymous mutation – A change in base sequence that alters the amino acid previously encoded there.

Nucleolus – An intra-nuclear organelle that is the site site of ribosomal RNA synthesis and ribosome assembly.

Nucleosome – The fundamental unit of chromatin, containing about 140 bp of DNA tightly associated with an octamer of histones (two each of four types).

Null allele – An allele that does not produce any functional product.

Okazaki fragments – Short pieces of newly synthesized DNA on the 3′-5′ side of a replication fork; that is, on the lagging strand.

Oncogene – A gene that can initiate a cancer, acting in a dominant manner.

Oxidative phosphorylation – The process by which mitochondria produce ATP concomitantly with generation of oxygen.

Pedigree – A pattern of inheritance of a phenotype in a family or larger group of related persons, usually summarized in a diagram with a set of standard symbols.

Pharmacogenomics – The study and applications of the genetic basis of individual variations in drug sensitivity, drug resistance, and drug metabolism.

Phenotype – The observable properties of a cell or an organism, which result from expression of the genotype and environmental influences.

Poly(A) – Polyadenylic acid, a simple nucleic acid sequence usually added at the 3′ end of a messenger RNA.

Polymerase – An enzyme that makes long polymers of nucleotides.

Polymorphism – A common (1% or greater) variant at a genetic locus.

Polyploidy – Presence of more than two complete sets of chromosomes in an embryo or cell.

Pre-implantation genetic diagnosis (PGD) – A test for disease-associated alleles on one cell taken from a very early-stage embryo produced by IVF.

Promoter – A region of DNA near the transcription start site of a gene, to which RNA polymerase and associated factors bind in order to initiate transcription.

Pronucleus – A gamete nucleus in a fertilized ovum, before they fuse to form the zygote nucleus.

Prophase – The first stage of mitosis or meiosis, during which the chromosomes begin to condense and the nuclear envelope disperses.

Proteome – The total set of proteins produced by an organism or a specific cell type.

Pseudoautosomal region – A small region of the X and Y chromosomes where they participate in recombination with each other. Genes in the pseudoautosomal region remain active on the inactive X chromosome.

Pseudogene – A gene copy that cannot produce a functional protein product. Pseudogenes arise from gene duplication or from retrotransposition of mRNAs.

Recessive allele – A form of a gene that does not produce a normal product, but will not affect an organism's phenotype if a normal allele is also present.

Recombinant DNA – DNA constructed in the laboratory from two or more segments that were not originally together. Recombinant DNAs are introduced into host organisms for research and commercial purposes.

Recombination – The process of exchange of DNA between synapsed chromatids at meiosis.

Recombination fraction – The frequency with which recombination (crossing-over) takes place between two flanking loci.

Regulatory sequence – A region of DNA associated with control of transcription of a nearby gene; includes promoters, enhancers, insulators, and so on.

Respiratory chain – Mitochondrial components that transfer electrons from high energy carriers (NADH and $FADH_2$) to oxygen, producing ATP from ADP and phosphate concurrently.

Retrotransposon – A DNA segment that can be multiplied by transcription, copying of mRNA into cDNA, and insertion of cDNA into a new genomic location, using enzymes encoded by itself.

Retrovirus – An RNA virus capable of being converted to DNA within a host cell and inserted into the host genome.

Reverse transcriptase – A type of polymerase that makes a DNA copy (cDNA) of an RNA molecule.

Ribose – The five-carbon sugar in ribonucleotides, the subunits of RNA.

Ribosome – The complex assemblage of RNA and protein to which mRNA binds and on which protein synthesis is carried out.

Robertsonian translocation – An end-to-end fusion of the long arms of two acrocentric chromosomes.

Short Interspersed Element (SINE) – One of the major types of dispersed repetitive sequences.

Single-nucleotide polymorphism (SNP) – A location in a DNA sequence where different individuals have a different base pair.

Sister chromatid exchange – Recombination between the two DNA strands of a single chromatid.

Stop codon – One of three trinucleotides (UAG, UAA, UGA) at which polypeptide synthesis will be terminated.

Synapsis – Precise side-by-side alignment of homologous chromosomes at meiotic prophase, prior to recombination.

Synonymous mutation – A base pair substitution that does not change the amino acid composition of the encoded protein.

Synteny – The relationship of genes on the same chromosome.

Tandem mass spectrometry – A technique that can measure the amount of many metabolites in a tiny sample of blood simultaneously; currently being applied to neo-natal genetic screening.

Tandem repeats – Multiple copies of a DNA sequence arranged in continuous linear order.

Telomere – The structure at the ends of chromosomes, consisting of tandem repeats of a hexanucleotide sequence with associated proteins.

Thymine – One of the pyrimidine bases in DNA; pairs with adenine.

Thyroglobulin – A large protein synthesized in thyroid follicular cells, which is iodinated on tyrosines and converted to thyroid hormone.

Thyroid stimulating hormone (TSH) – A polypeptide synthesized by the anterior pituitary, which regulates the synthesis of thyroid hormone by the thyroid gland.

Topoisomerase – An enzyme required for DNA replication; it makes single-strand breaks, which allows correction of either underwinding or overwinding that would interfere with replication, then seals the breaks.

Transcription – Synthesis of RNA from a DNA template.

Transcriptome – The complete set of genes expressed in a specific cell type.

Transfer RNA (tRNA) – Small RNAs to which amino acids are attached, prior to incorporation into a polypeptide.

Transgene – A gene transferred into a host organism by molecular methods.

Translation – Synthesis of a polypeptide on a ribosome from the sequence information in mRNA.

Translocase – A protein that transports a specific class of small molecules from one side of a membrane to the other.

Transmission genetics – The branch of genetics that deals with patterns of inheritance from one generation to another.

Triplet repeat disease – A disease that results from extensive multiplication of a trinucleotide in a gene, usually leading to a protein that functions abnormally.

Trisomy – Presence of three copies of one type of chromosome in otherwise diploid cells.

Tumor suppressor gene – A gene whose normal function is to prevent proliferation of cells. When both copies are inactivated or lost, uncontrolled cell division ensues.

Uniparental disomy – Origin of both copies of a specific chromosome from one parent, with none from the other parent.

Uracil – A base found in RNA, wherever thymine occurred in the corresponding DNA.

Variable expressivity – Variable phenotypes caused by the same mutant allele in different persons, usually in response to polymorphisms in other genes.

Index